The Elements of
Electronic
Communication

The Elements of Electronic Communication

Heidi Schultz

The Kenan-Flagler Business School
The University of North Carolina at Chapel Hill

Allyn and Bacon

Boston   London   Toronto   Sydney   Tokyo   Singapore

Vice President: Eben W. Ludlow
Editorial Assistant: Grace Trudo
Executive Marketing Manager: Lisa Kimball
Editorial-Production Service: Communicáto, Ltd.
Text Design and Electronic Composition: Denise Hoffman
Composition Buyer: Linda Cox
Manufacturing Buyer: Suzanne Lareau
Cover Administrator: Jenny Hart

Between the time web site information is gathered and then published, it is not unusual for some sites to have closed. Also, the transcription of URLs can result in unintended typographical errors. The publisher would appreciate being notified of any problems with URLs so that they may be corrected in subsequent editions.

Many of the designations used by manufacturers and sellers to distinguish their products are claimed as trademarks. Where those designations appear in this book and Allyn and Bacon was aware of a trademark claim, the designations have been printed in caps or initial caps.

Library of Congress Cataloging-in-Publication Data

Schultz, Heidi Maria.
    The elements of electronic communication / Heidi Schultz.
        p.   cm. — (The elements of composition series)
      ISBN 0–205–28646–1
      1. English language—Business English—Data processing.   2. World Wide Web (Information retrieval system).   3. Business communication—Data processing.   4. Business writing—Data processing.   5. Usenet (Computer network).   6. Electronic mail systems.   I. Title.   II. Series.
PE1479.B87S38   1999
808'.066650'0285—dc21   99-43251
                                            CIP

Printed in the United States of America
10 9 8 7 6 5 4 3 2 1   04 03 02 01 00 99

Permissions Credits
Figures 1.1, 1.2, 1.3, and 1.5 (pp. 35, 40, 44–45, and 57, respectively): Template from Microsoft Office v. 4.0 (Seattle: Microsoft, 1998).
Figure 1.3 (pp. 44–45): Text reprinted with permission from Scott Blalock.
U.S. Post Office logo (p. 97): Unifying header from the United States Postal Service web site, used with permission of the United States Postal Service, Office of Corporate Identity.

For my husband, Adam,
who takes great pleasure
in the verbal acrobatics
associated with all communication

Contents

2 The World Wide Web   63

Preface

It's long been said that an infinite number of monkeys, pounding away at an infinite number of typewriters, would eventually recreate the works of William Shakespeare. Thanks to the Internet, we now know that this is not true.

—California First Amendment Coalition et al. v. Calderon,
Slip Paper Option, 28 April 1998, quoting
Brandsburg v. Hayes, 480 U.S. 684 (1972)

This image of the Internet as a haven for poor communication is fairly widespread. And even though it's probably safe to assume most of us don't aspire to the literary greatness of William Shakespeare, we're certainly faced more and more with communicating electronically. Everyone with a computer and a connection to the Internet has the *technical capability* to communicate electronically. Of course, this doesn't mean those same folks have the *rhetorical ability* to do so effectively. As Internet technology progresses at a brisk pace, electronic communication gets pulled along rather quickly; however, the *conventions* for doing so follow more slowly. As a result, many people adopt an "anything goes" style for email messages, web site text, and newsgroup postings. Moreover, without rules governing effective electronic communication, inadequate communication skills not only thrive but also breed and spread.

Okay. Okay. Maybe comparing deficient communication skills to a pervasive, communicable disease is a bit overly dramatic. So let me put it this way: The purpose of any communication—whether electronic, spoken, or written (in the traditional sense of the word)—is to get results. Otherwise, the communication is a waste of time for both the writer and reader. If we can agree on this, then we can also agree on the following statement:

If you want your electronic messages to get the results you desire, you'll need to adopt and practice effective communication strategies.

For just a moment, consider the following questions and your answers to them:

- Do you want your employer to adopt your proposal?
- Do you want your instructor to change your grade?
- Do you want your legislator to bring your idea to the Senate floor?
- Do you want to find out about the best time to plant and fertilize tulip bulbs?
- Do you want to sell your craft kits?

If you answered yes to any of these questions (or questions that are similar to these) and if you use or intend to use the Internet to communicate, then you've just established the importance of effective electronic communication.

In *The Elements of Electronic Communication,* I don't spend much time discussing technical issues related to software and hardware. For this information, simply visit your local library, spend some time with the user manuals for your equipment, or contact the technology information specialist in your organization. I will, however, introduce you to strategies that make email messages, web site text, and newsgroup postings effective. To this end, you'll learn how word choice can affect meaning, how organization can affect clarity, and how typographical techniques can emphasize ideas. With a little reading and a lot of practice, you'll be creating effective electronic messages sooner than you might think!

Acknowledgments

Although I've spent many solitary hours at my computer working on this book, I know that any writing project is truly collaborative. And for the help I received with this project, I'm grateful.

Specifically, I thank Eben Ludlow, my editor at Allyn and Bacon, for the excellent editorial suggestions he provided throughout the process, and his editorial assistants, Linda D'Angelo, Tania Sanchez, and Grace

Trudo. Thanks also to the colleagues who reviewed my manuscript for Allyn and Bacon and made insightful suggestions: Sam Dragga, Texas Tech University; Ed Klonoski, Charter Oak State College; Nancy O'Rourke, Utah State University; and Philip Vassallo, Middlesex County College.

The technology support staff at UNC–Chapel Hill's Kenan-Flagler Business School answered many of my technological questions. They were an invaluable resource to me.

I thank my parents, Hans and Helga, and my sister, Helen, who provided me with love and encouragement. And finally, I thank my husband, Adam, who read my drafts, offered wonderful ideas, and cooked many a dinner as the deadline approached.

Introduction

Let me ask you something: Why should you care about electronic communication? After all, for most folks, electronic communication is as easy as opening an email program or joining a newsgroup discussion and clicking away on the keyboard, right? If it were only that simple, we wouldn't be bombarded with so many poorly organized email messages, difficult to navigate web sites, and truly offensive newsgroup discussions.

Within the last decade, electronic communication has become the efficient communication equalizer. Everyone seems to have electronic access to everyone else, significantly increasing the pace of communication. Because of this breakneck speed, many electronic communicators fail to consider the consequences of exchanging hastily composed and immediately available messages. For instance, a junior consultant in a regional consulting agency sent an email message to the company's president, proposing several consulting strategies that the president decided to adopt. When the president sent a companywide email announcing the changes, two things happened:

- The junior consultant's boss was offended that his employee had gone over his head.
- Most of the agency's other consultants were offended that they weren't involved in a discussion about the proposed changes.

As this example indicates, electronic communication has also caused traditional communication boundaries to crumble. Suppose you need to communicate with your company's CEO or your hospital's chief administrator. Why make an appointment? Send an email message to convey your agenda. Are you interested in advice on lawyer's fees? No need to schedule a meeting. Just join a newsgroup to solicit information on the subject. Do you want to sell your new invention? Don't make appointments with distributors. Create a web site advertising your product.

Given the accessibility of your intended audience and the speed with which you are able to blast your electronic messages into cyberspace, it's important that you know how to write for today's electronic environment. Let me stress that the key word here is *write*. Once your home, office, or school computer has been connected to the Internet, you've mastered the technology (or someone has mastered it for you). That's great, but that's not enough. You also need to use efficient writing strategies for your electronic messages to get the results you desire.

Organization of This Book

In Chapters 1 and 2 on email and the World Wide Web, you'll learn how to analyze your audience, determine your message's purpose, adopt an appropriate style and tone, and create an effectively organized message. You'll also learn about the very public nature of all electronic communication and discover which important legal and ethical issues should concern you. Chapter 3 on the Usenet provides suggestions for appropriate newsgroup communication, considering not only a message's organization but also its content. Moreover, this chapter provides compelling reasons as to why businesses, educational institutions, government agencies, and health care organizations are interested in posting and reading Usenet messages.

To help you untangle some of the specialized terms associated with electronic communication, I've placed a glossary just after this introduction entitled Guide to Specialized Terms for Electronic Communication. While you'll usually find such information at the end of most books, I decided to put it close to the beginning for two reasons:

1. For this guide to be helpful, I need to be sure you know it's there.
2. Before reading the chapters, you may decide to review this information briefly.

Note: If you'd like some help understanding cyberspace—the domain of email, the World Wide Web, and Usenet—you should finish reading this brief introduction. If, however, you have a solid conceptual grasp of the subject, you're welcome to skip ahead to the Guide to Specialized Terms or to the chapters.

Using the Internet

To communicate electronically—whether by email message, web site, or Usenet—you'll need to establish a connection to the Internet. The Internet is an expansive network of linked computers that can communicate words, text, graphics, and audio *throughout the world,* to and from homes, businesses, educational institutions, and government organizations. There are also systems of linked computers called *intranets.* These systems also communicate words, text, graphics, and audio, but they do so *within a given organization,* such as a business, university, or government agency. Computer scientists developed intranets to help companies ensure privacy and keep outsiders from accessing proprietary information. It may help to think of the intranet as a closed system and the Internet as an open system. Keep in mind, however, that many of the rules governing effective electronic communication apply to both systems.

Figure I.1 (page 4) demonstrates how intranets share information. As you can see, communication exchange occurs in an unrestricted manner among computers within a specific organization. Communication also can pass outside the organization's system. However, communication that wants to enter the organization is severely restricted.

Figure I.2 (page 5) offers a schematic diagram of the Internet within the context of personal and organizational computers. Note that although a firewall isn't shown for the Educational Institution Network, certain databases within any organization can be protected this way if the administration so chooses.

FIGURE I.1   An Intranet

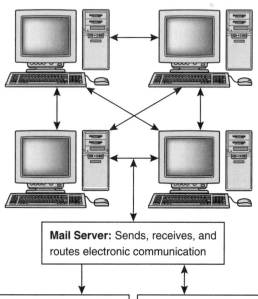

Mail Server: Sends, receives, and routes electronic communication

Company Database: Keeps copies of electronic files—all messages anyone sends and receives and all web sites anyone visits

Server: Sends electronic communication out to interested parties and allows approved information to flow in; doesn't allow access to users outside the organization

Firewall: Protects company information from users outside the organization; information can flow from the organization to the outside, but outside users can't gain electronic access

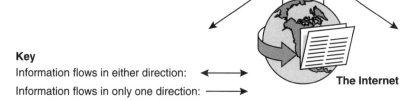

Key
Information flows in either direction:
Information flows in only one direction:

The Internet

   Connecting to an intranet is relatively easy because it's part of an organization's internal communication system. In this situation, technology systems administrators will typically connect your computer to this

FIGURE I.2   The Internet

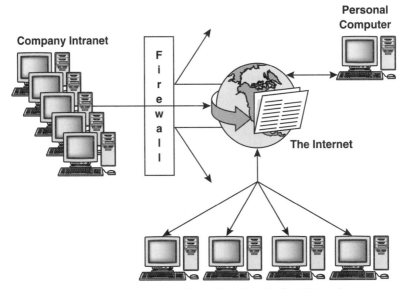

Key
Information flows in either direction: ◄────►
Information flows in only one direction: ────►

closed communication system, thereby limiting your involvement in technological concerns. For the same reasons, connecting to the Internet can be relatively simple if you're also accessing it from your organization's hardware. Many organizations and educational institutions offer employees and students direct connections to the Internet through high-speed cables, again set up by systems administrators.

Connecting to the Internet from a personal account, however, can be a bit trickier. In most cases, these decisions are involved:

- Deciding whether to connect through a telephone line, a satellite service, or even a cellular telephone requires time-consuming research.

- Choosing high-speed connection options, which carry acronyms such as ISDN and ADSL, can be perplexing.

- Selecting an ethical and dependable Internet Service Provider—whether a telephone, electric, or cable television company—to establish an Internet account also demands some attention.

Figure I.3 offers some basic options for gaining access to the Internet.

So, what's the good news here? Making all these decisions seems like a lot of work. But once you have Internet access, you'll be able to do several extraordinary things:

- Send and receive email messages, making your communication more efficient

---

FIGURE I.3   Accessing the Internet and What You'll Find Once You Get There

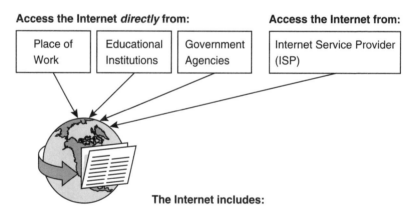

Access the Internet *directly* from:

Place of Work	Educational Institutions	Government Agencies

Access the Internet from:

Internet Service Provider (ISP)

The Internet includes:

- **Electronic Mail:** Programs that send and receive messages over a computer network. You can send email to or receive email from anyone on the Internet. You can also attach files, such as word-processed documents, to email messages.

- **World Wide Web (WWW):** A hypertext-based system that allows you to navigate the Internet. Because of its versatility, the WWW is the fastest-growing component of the Internet. It can include audio from radio stations, video from television news networks, web pages, and other multimedia information.

- **Usenet:** A network that provides access to electronic discussion groups, which usually have a focus, such as gardening or auto racing.

- Add your own web site to the World Wide Web, offering your product or service to the world

- Post messages on Usenet newsgroups, soliciting or offering advice on questions related to educational, legal, and medical issues, for example

What's the other good news? Now that you understand what it takes to establish an Internet connection and how Internet users share information, you're ready to begin creating effective electronic messages. So, let's get started!

Guide to
Specialized Terms for
Electronic Communication

To guide your understanding of the discipline-specific language associated with electronic communication, this guide includes the definitions of technical terms, Internet jargon, and cyberslang.

Alias is an easy-to-remember version of an alphanumeric email address.

Alphanumeric email address is a combination of letters and numbers assigned to you by your system administrator. The address *hms106@ mindspring.com* is an example of an alphanumeric email address.

Angle brackets (< >) sometimes surround email or web addresses to indicate that the information contained within the brackets is a single unit. Using angle brackets in either handwritten or printed text minimizes misinterpretation of your address by your readers. For example, *<http://www.worldwildlife.org>* is a single web address as is *<Heidi_ Schultz@unc.edu>*. However, don't use angle brackets when keying in a web address or email address with a browser. Using brackets is simply a convention used within text. It's also important to note that certain email and word-processing programs automatically recognize web and email addresses and convert them to a smaller, underlined font, like this: Helen_Jona@mindspring.com. If your program has this function, you can safely omit the brackets.

ASCII stands for *American Standard Code for Information Interchange,* which is the most basic format for transferring files in unformatted text versions among computers.

Asynchronous communication refers to the delayed communication typically associated with email. In other words, the participants in an electronic conversation don't communicate with one another in real time. When you send a letter through the U.S. mail and wait for a reply, you are also participating in asynchronous communication. (See also **Synchronous communication.**)

At or **@** is the symbol that separates the username from the domain name in email addresses, indicating you are *at* that particular address.

Attachments are electronic files or documents connected to other electronic files for the purpose of transferring those files. For example, you can attach a Microsoft Word document to an email message when you want to forward that document electronically.

Bookmark lists, also called *hotlists* or *favorites,* contain lists of links to your favorite web sites.

Bookmarks allow you to access favorite web sites quickly.

Browsers are software programs that allow you to navigate the Internet. Netscape Now is a graphics-based World Wide Web browser that allows you to maneuver around the web and download particular web sites. Other graphic browsers include HotJava, Internet Explorer, and Mosaic. Lynx is the most popular text-only browser.

Bulletin Board System (BBS) is a computer system that serves as a forum for a particular group's interest, such as endangered animals.

Bulletin boards allow you to leave electronic messages or share news. Anyone can access, read, and respond to the messages you leave there. Be aware, also, that marketers and other interested parties can get your email address and send you large numbers of unsolicited messages if you post a message to a bulletin board or newsgroup. (See also **Newsgroups.**)

Chat room is an electronic place for synchronous, or real-time, online conversations. (See also **Synchronous communication.**)

Chatting is a way for people to communicate online in real time by typing messages to one another.

Client is the software application that requests information from a server. In other words, it brings information back from the Internet to your computer. When you surf the Internet, you can think of yourself, your computer, or your browser as the Internet client.

Cookie is a file fed to your computer when you visit a certain web site. Should you return to the same site, the cookie file will allow the web site to identify you as a return visitor.

Cursor is the blinking vertical line that follows the characters you type in an email message or newsgroup posting. On a web page, the cursor is the small arrow that follows the mouse and shows you where you are on the screen. When you place the cursor over text or icons that are links, it changes from an arrow to a pointing finger.

Cyber- is a prefix that describes something electronic. For example, *cybercommerce* refers to buying and selling products and services over the World Wide Web. *Cyberlots* are the electronic equivalent of car lots.

Cyberspace is a synonym for *Internet* coined by William Gibson in his novel *Neuromancer* that's used to describe the Internet and other online environments. It's typical to hear that someone is "navigating through cyberspace" when in an online environment.

Domain name is the address used for identifying and locating a given computer connected to the Internet.

Domain Name System (DNS) is a database that translates domain names, or web addresses, into language that computers can understand.

Dot or **.** is the symbol that separates the elements of email addresses, web addresses, and newsgroup addresses.

Download refers to the transfer of files, such as web sites and software, from a distant computer to your computer. When you access a web site and it appears on your computer screen, you download the information.

Email stands for *electronic mail*. It is provided by a software program capable of sending and receiving messages over an electronic network. Popular email programs include Outlook and DaVinci.

Email address is the address that you use to send and receive email. It consists of your username, the @ symbol, and the domain name. For example, my email address is *<Heidi_Schultz@unc.edu>*. (Remember, the angle brackets are not a part of the address.)

Emoticons (or **emotion icons**) are the strings of characters found in email messages that serve as substitutes for facial expressions and body language. To avoid misinterpretation, use them carefully and only in informal messages. Also known as *smileys*.

Encryption software allows you to protect files as you transmit and receive information over the Internet. By encrypting files, you prevent others from accessing personal information, such as your credit card number or sensitive medical or financial data. (See also **Private encryption** and **Public encryption**.)

FAQ is the acronym for *Frequently Asked Question*. FAQs list and answer the most common questions users might have about particular subjects typically found on web sites or within newsgroups.

Filter is a software program that allows you to block access to certain web sites. It's useful for parents who want to limit their children's access to inappropriate information.

Firewall is the boundary that prevents Internet traffic from crossing from one system to another. Companies protect proprietary information on their intranets with firewalls.

Flaming is the electronic equivalent of a tongue-lashing, usually conveyed in all uppercase letters.

Format negotiation refers to converting data such as text, image, and/or sound formats by a server into a format that the requesting client can understand. It's useful because it allows access by a client to a wider range of web sites.

Frames are the separate windows within a web site's main window.

FTP is the acronym for *File Transfer Protocol*, which is a set of commands that transfers files between computers linked via the Internet. But as format negotiation and attachments become more popular, FTP is becoming obsolete.

GIF image, which stands for *Graphics Interchange Format*, is one of two common formats for graphics or image files in web documents. (See also **JPEG**.)

Gopher is a system that allows you to access information on the Internet.

Graphic web site offers all the textual, video, audio, and sound files to a visitor.

Hardware refers to the monitor, keyboard, mouse, speakers, and central processing unit (CPU). It's the computer—the machine and its parts.

Hit refers to the number of items a search engine returns based on an inquiry from your browser. It can also refer to the number of times a particular web site has been visited. (See also **Search engines.**)

Home page is the first page of a web site. It's the hypertext page with the index, introductory information, and links to other pages either within the same site or other sites. Individuals and organizations have home pages.

Host is a computer directly connected to the Internet.

HTML is the acronym for *Hypertext Markup Language.* It's a computer code that allows you to create a site and pages for the World Wide Web. HTML is a special way to tag files that web browsers can read.

HTTP is the acronym for *Hypertext Transport Protocol,* a system of rules for moving HTML documents across the web.

Hyperlinks are highlighted or underlined text or graphics on web pages that allow users to jump between two locations on the web. They are sometimes known as *hotlinks.* You know you're on a link when the cursor changes from an arrow to an icon that looks like a pointing finger. (See also **Links.**)

Hypermedia refers to integrated text, graphics, audio, and video tied to a hypertext system.

Hypertext is a document written or coded in HTML. (See also **HTML.**)

Internet is an expansive computer network that offers access to the World Wide Web, email, and the Usenet.

Internet Explorer is the name of a popular browser that accesses information on the Internet.

Internet service provider (ISP) provides access to the Internet, similar to how a phone company provides access to telephone lines. AT&T Worldnet, GTE.Net, and Mindspring, for example, are all ISPs.

Intranet is an internal web that organizations use to exchange internal, private communication. Intranets allow employees to exchange confidential school- or work-related information without allowing the public access to that information.

IRC is the acronym for *Internet Relay Chat.* It's a large, unregulated chat system for instantaneous communication.

JPEG, which stands for *Joint Photographic Experts Group,* is one of two common formats for graphics or image files in web documents. (See also **GIF.**)

Junk email is unsolicited commercial email also known as *spam.*

Keywords are the words you enter into a search engine to locate specific information on the World Wide Web. (See also **Search engines.**)

Links are the highlighted words, phrases, and graphics on a web site that allow you to connect or jump to other places in the same or different web sites. Internal links allow you to jump among pages within a specific web site. External links allow you to jump among many web sites (see also **Hyperlinks**).

Listserv is an online service that allows participants to discuss similar interests. Participants subscribe and unsubscribe to listservs via a central service.

Lurkers are people who read newsgroup postings or listserv postings without participating in online discussions.

Lynx is a popular text-based Internet browser. As such, it doesn't convey graphics or video.

Modem, which is short for *Modulator Demodulator,* is the internal or external device on your computer that connects it to the Internet by means of a phone line.

Monitor is someone who electronically moderates a listserv discussion. (It's also the electronic equipment, or computer screen, that displays web sites, email messages, and newsgroup discussions.)

Mosaic is an early graphic web browser.

Multimedia stands for the various media—such as text, graphics, audio, and video—incorporated in electronic documents, programs, and products.

Navigation refers to the movement among pages within a web site and among many web sites. By simply clicking on links, you can move from place to place on the Internet—that is, you can *navigate* the Internet.

Navigational bar refers to a collection of topics (text only) or pictures with topics (text and graphics) that reveal links to other pages. They often appear within frames, like these:

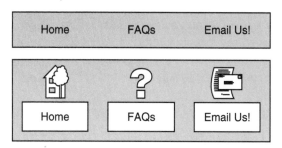

Netiquette refers to appropriate communication behavior on the Internet. The term was coined from *net* and *etiquette*.

Netscape Now is a popular browser for accessing the Internet.

Newsgroups are groups of people and their collective postings on Usenet networks. Each newsgroup has a specific topic or focus, and anyone can participate by simply subscribing. (See also **Bulletin boards.**)

Online refers to being on the Internet or some other electronic network.

Password is the personal code that allows you to access your Internet service provider (ISP) account. You'll be asked to create a password for some web sites you visit. It's your protection against someone using your account without your permission. The best passwords combine letters and numbers and don't reflect anything personal, such as your birthdate, license plate number, address, or telephone number. It's important to keep your password confidential. Also, because you'll accumulate numerous passwords over time, you should keep a list of them in a safe place.

Posting refers to creating an online message. When you send a message to a newsgroup, you *post* it.

Privacy policy is a statement on a web site that addresses how the site and its owner(s) will use information gathered about you. A responsible site displays its privacy policy prominently.

Private encryption is a way to send private information through the Internet in which both the sender and receiver have a code to scramble and unscramble messages. For example, by scrambling the information before sending it to an online retailer, you can protect your private information—credit card number or social security number, for example—from getting into the wrong hands.

Public encryption is another way to protect the information you send through the Internet. Public encryption software allows only one person involved in the transmission process to unscramble the message, whereas private encryption software allows folks on both ends to scramble and unscramble information.

Registration is the process through which you or your organization receives a domain name.

Search engines are programs such as AltaVista and Yahoo! that work with your browser to find information on the web. They're sometimes called *search tools.*

Server is the main computer that handles requests from clients for data, email, file transfer, and other network services. Think of it this way: A server *serves up* information. A server can also be an application that makes data available on a network.

Signature is a file of several lines appended to the end of an email or Usenet message. It can include the sender's name, position, email address, company or school address, and other information, such as quotations and personal philosophies. You should eliminate extraneous information to keep your signature short.

Sites are specific places on the web that are also known as *web sites.*

Snail mail refers to mail that's delivered by humans rather than by machines. The U.S. Postal Service, Federal Express, and United Parcel Service all deliver snail mail. It's so named because compared to email, it's relatively slow.

Software refers to the programs that actually run your computer. When you insert a CD or floppy disk into your computer, it reads the information and then stores it for future use. Software is not part of the computer until you put it into the computer's memory.

Spam refers to the unsolicited electronic mailings that usually promote shady dealings. If you don't indicate that your spam is actually an advertisement (which, by the way, you need to do in the subject line), the U.S. government can fine you up to $11,000. Being *spammed* is the act of receiving spam, a *spammer* is someone who sends spam, and *spamming* is the act of sending spam.

Subject directory is the list of web sites generated by using a search engine. (See also **Search engines.**)

Subscribe refers to setting up your computer so you can send and receive messages for a particular newsgroup.

Surfing is what you do when you explore the Internet. A *surfer* is someone who explores, or *surfs,* the Internet.

Synchronous communication refers to real-time communication among people who can communicate simultaneously via the Internet. (See also **Asynchronous communication.**)

System operator or **System administrator** is the person in charge of overseeing Internet use in a particular organization. This person also has the ability to monitor all content. (See also **Monitor.**)

Telnet is a method for connecting to a remote computer by using a username and a password.

Text-only web site is devoid of graphics, audio, and visuals.

Thread is a sequence of related postings on a specific subject.

URL is the acronym for *Uniform Resource Locator,* which is the specific web address that lets you locate a particular site. It was developed to *locate* a site on the Web according to a *uniformed* format.

Usenet is a network that provides access to an electronic newsgroup to exchange messages on a given topic.

Username is the information that, along with your password, lets you access your computer account. Your Internet email address most likely begins with your username. My username is *Heidi_Schultz.*

Virtual is a synonym for *online. Virtual communication* is the same thing as *online communication,* and *virtual reality* is the same thing as *online reality.*

Virus is a destructive or annoying file planted in a computer that can corrupt files and damage systems.

Wallpaper is the background you see on a web site. It may be as simple as a white background or as complex as a company's name superimposed throughout.

Web browser, also known as a *browser,* is a program for navigating the Internet. Most browsers display attractively formatted pages and graphics and let you click on hyperlinks to jump from one web page to another. Browsers can be graphic browsers or text-only browsers. Widely used graphic browsers include HotJava, Microsoft Explorer, NCSA Mosaic, and Netscape Navigator. A popular text-only browser is Lynx.

Web pages include text, pictures, audio, and video. Web pages can be linked to other pages through hyperlinks. All the linked web sites make up the World Wide Web. Web pages can be subsets of a specific web site or several web sites. (See also **Hyperlinks.**)

Web site is the specific location on the web made up of several pages. Think of a web site as a book and the pages within as chapters of that book.

Window is a specific area on your computer screen that presents information within a defined space.

Wired is what you are when you're online.

World Wide Web (WWW) is a major part of the Internet that connects hypertext data—including text, graphics, and audio—among smaller networks and personal computers. (See also **Hypertext.**)

1

Electronic Mail

During the past 15 years or so, the number of email users has increased 100-fold! Some 35 million of us spend much of our time sitting at computers, pounding out more than 100 million email messages every day. Our computers are linked through the Internet or intranets, which allow us to send and receive urgent business messages, convey important medical and legislative information, and simply stay in touch with scattered friends and relatives around the community, the country, and the world.

The widespread use of email has changed much about the way we communicate, breaking social hierarchies, geographical boundaries, and time barriers. Email also has challenged the traditional channels of communication among members within organizations. Closed doors no longer translate into inaccessible CEOs, journalists, lawyers, physicians, professors, teachers, or upper management. Simply put, everyone with email has access to everyone else with email—and that's a lot of accessibility! People who live or work down the hall, up the street, across campus, or over the continental divide are as reachable as those folks who share your school, your office, or your house. Communication with your Japanese relatives, South African business partners, or Norwegian academic contacts is no longer restricted by time zone considerations. You'd never call someone in Europe at 6:00 p.m. mountain time. But with email, any time is the right time. Given the convenience of email, we have once again become prolific message writers, which may seem like a blessing—until you open your email to discover 50 messages that require your attention!

Because email sidesteps traditional communication channels, it is primarily an informal communication tool. As does its accessibility, email's informality brings challenges to electronic communication. It has an interesting way of getting people to drop their inhibitions because of a *perceived* anonymity some writers tend to adopt. Without the accountability that comes with face-to-face conversations, email users sometimes write what they would never say to a recipient's face or write in a formal letter. For example, a manager from a large organization wrote the following email message to a technology-support staff person, who had been busily exchanging the organization's 15-inch monitors for 17-inch monitors:

> NEVER, NEVER touch my machine without telling me first! At best, it's rude. At worst, it's unprofessional!

Thinking he was doing the manager a favor by exchanging the smaller monitor for a larger one, the technology-support person said, "I couldn't believe it. She's always been real nice to my face. But her reaction was a surprise, if I've ever had one in this job."

Because it's so easy to send email messages, we send lots and lots of them. An article published in *The Baltimore Sun* revealed that President Clinton received 2.7 million email messages from 1993 to 1998, when he started to get 2,500 email messages every day. And a recent survey revealed that email messages interrupt the typical worker three times every hour to respond to and send an average of 190 messages each day.

This voluminous exchange of information creates yet another challenge with email. Connecting yourself to email—whether at work, at home, or at school—can lead to something called *communication enslavement,* in which you feel as though you're always on call because email writers expect instant responses to their messages. Moreover, email's quick pace causes many people to check messages regularly or have email programs automatically alert them by beeping each time a new message arrives. (In fact, as I was writing the sentence about beeping, *my* email beeped, indicating a new message had arrived! Despite the fast-approaching deadline for this book, I stopped writing to check the message—but it wasn't anything important.) While curiosity is a powerful emotion, I have no doubt that continual email interruptions can reduce anyone's productivity.

Given these challenges, you'll want to ask and answer some questions before sending any email message. So let's do that now.

Why Send Email?

Despite its challenges, email certainly has advantages. Once you've written your message, you don't have to address a separate envelope, seal the message in the envelope, put a stamp on the envelope, and drop it in a mailbox. Furthermore, you can send the same message to lots of people without taking time to make photocopies. With a few simple keystrokes, you can send your message to 1 or 100 people. So, email is extremely convenient.

Email is also the fastest way to send a message. When you send an email to someone in your organization, your message stays within your organization's mail server. Because your electronic message doesn't have to travel on outside servers, your message is transmitted in seconds. If, however, you send an email message outside your organization—for example, across the country or halfway around the world—your message must move from one server to another. While this process slows your message's transmission, 99% of the time, your message arrives in less than a minute.

Email is also inexpensive, when you consider paper, envelope, and postage expenses. Sure, messages travel over expensive computer networks. But because those networks serve additional purposes, we can't blame those expenses exclusively on email.

When Is Email Appropriate?

Even though email is convenient, fast, and inexpensive, you'll need to determine its appropriateness in relation to your specific communication situation. Email is appropriate in the following instances:

- When you don't need to speak to someone face to face
- When the recipient of your message is not immediately available
- When the information you're sending is not sensitive or personal

Email is also appropriate when you need to convey detailed information that would be cumbersome to deliver over the phone, on voice mail, or even in a face-to-face conversation. For example, when your message includes names and addresses or long account and telephone numbers, email allows you to deliver that information not only efficiently but also accurately. One word of caution: If you are sending account numbers, credit card numbers, or social security numbers, you need to be certain that the information is protected from unintended eyes. Knowing *when* to use email instead of another communication medium is as important as what information is contained in the message.

When Is Email Inappropriate?

Never write or send an email message when the information you are conveying is personal, confidential, or inflammatory. Simply stated, email is *not* private. Even if you delete personal, confidential, or inflammatory email messages from your computer, they may still be on your server's system indefinitely. Internet service providers, government agencies, educational institutions, and corporations all maintain copies of your messages long after you delete them from your email account. For instance, the now-infamous *Starr Report* revealed that Monica Lewinsky's "deleted email [had been] recovered from her home computer and her Pentagon computer." In October 1998, at the start of the Department of Justice's antitrust trial against software giant Microsoft Corporation, government lawyers read dozens of CEO Bill Gates's private email messages. In both cases, the information conveyed in email messages was personal, confidential, inflammatory, or some combination of all three.

Here's another example: In late 1998, the U.S. commander overseeing the no-fly zone in northern Iraq sent the following inflammatory email message:

It's a good day for bombing.

While Air Force Brigadier General David Deptuala admitted he should have chosen his words more carefully, he learned a very important lesson the hard way: that the air force's systems administrator has the technological ability to monitor all email activity.

What does all this mean? It means you should never, *never* write anything in an email message at home, at work, or at school that you wouldn't write on the back of a postcard, want to see on a billboard, or be ashamed to show your Great Aunt Elly.

Email is also inappropriate in the following situations:

- When facial expressions, gestures, or speech patterns are important
- When a message requires your actual signature
- When you need to send mass mailings but fail to limit your recipients to a specific target audience

Email Accounts and Addresses

To send or receive an email message, you'll need an email account provided by your educational institution, your place of work, or an Internet service provider. If you get an Internet service provider account, it will cost you under $20.00 per month.

You'll get an email address in conjunction with your account. An email address consists of two parts joined by the @ sign. The first part is called the *username* and is located to the left of the @ sign. You'll sometimes hear the username referred to as the *account name* or the *userid* (user i.d.), where *id* stands for *identification.*

When opening an email account, if you have the option, choose as your username your personal name. Your name attached to the rest of your email address will help identify you. This consideration is increasingly important for two reasons:

1. More and more telephone companies are starting to include personal email addresses in phone books. The first such service was offered by Concord Telephone of North Carolina in October 1998. If the purpose of a published email address is to make you accessible, then make your address easy for folks to remember.

2. Alphanumeric email addresses, such as *sjh986@company.com,* are too cryptic to identify you.

If, however, you decide to choose a name other than your personal name, be professional. Usernames such as *FastEddie, Poohbear, Hulkster, HotMama,* and *SweetBaby* are only appropriate in the most intimate of relationships. They don't translate well to the professional or business world. In addition to the unprofessional image that these usernames convey, it's cumbersome to change your email address once you've chosen a username and distributed your email address because you then have to notify everyone of the change. If you don't like filling out those "change of address" cards at the post office, calling magazines, and dropping by the bank and the Department of Motor Vehicles, you won't like changing your email address. The bottom line: Choose a username that's appropriate for as many contexts—professional and personal—as possible.

With some Internet service providers, however, you won't have the option to choose your username. Rather, the Internet service provider will assign you an alphanumeric email address that might consist of your initials and some numbers, such as *hm12029@network.com.* Because this kind of address is difficult to remember and too cryptic to identify you in the "From" line, consider adopting an alias. An *alias* is an electronic pseudonym—usually, your actual name—that offers an easily identifiable email address. Instead of keying in *hm12029@network.com* every time you want to send that person a message, you simply key in *Harvey_Miller@network.com.* With an alias, the server forwards messages sent to *Harvey_Miller@network.com* to the appropriate mailbox at *hm12029@network.com.* While this procedure may sound complicated, it's really not. All email software programs, even primitive versions, allow you to adopt an alias.

So much for the information to the left of the @ sign. The information that appears to the right of the @ sign is called the *domain name.* It has at least two parts separated with a period: (1) a specific organization's server and (2) the kind of organization that operates the email server. Look at my email address, for example:

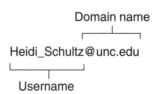

As you can see, my specific organization's server is *unc,* or *The University of North Carolina at Chapel Hill,* and *.edu* indicates that the organization is an *educational* institution. In the email address *Jeff_Austin@anybank. com,* the organization's server is a specific bank and *.com* indicates the organization is, for example, a commercial lending institution. Common domain suffixes include the following:

.com	**com**mercial
.edu	**edu**cation
.gov	**gov**ernment
.mil	**mil**itary
.net	**net**work management or Inter**net** service provider
.org	noncommercial/nonprofit **org**anization

Outside the United States, a two-letter suffix refers to a specific country. For example, when the domain name ends in *.ca,* the country referred to is *Canada.* When the domain name ends in *.de,* the country referred to is *Germany.*

Conducting a Situational Analysis

When you decide to communicate via electronic mail, you'll need to consider more than the advantages, limitations, and appropriateness of sending email messages. In addition, you should consider these points:

1. Have a well-defined purpose for each message, and consider the most effective way to organize that message.
2. Identify your readers' characteristics.
3. Carefully consider your relationship to your readers.

Taken together, these considerations guide you through a *situational analysis,* a method for constructing effective messages in *any* communication medium.

   Roman Jakobson, a respected rhetorician, devised the *communication triangle* to help communicators—whether writing or speaking—complete a thorough situational analysis. Using Jakobson's communication triangle makes any rhetorical situation easier to untangle. Here's how it works: For any message—an email message, a business letter, an oral presentation, a web site, or a newsgroup message—consider your role, your audience, and your message:

My Role
What's my status?
What's my affiliation?
What's my relationship to my audience?

My Message
How do I define my message's purpose?
Do I want to inform? entertain? persuade?

My Audience
Who is my audience?
What do I want my audience to think?
What do I want my audience to do?

Now that we have an understanding of the three components of an email message—purpose, audience, and writer's role—let's explore each in more detail.

Writer's Role

Because your relationship to your audience affects your choice of words and the way you organize your documents, it's important that you have a clear understanding of *your role* in the communication process. For example, you may be in the role of subordinate if you're writing to your supervisor. In that case, you may start your email message with a salutation followed by your supervisor's name and title. You may also buffer your ideas with hedges such as *I believe* and *I think*. If, however, you're writing an email message to a close friend, you'll probably forgo the salutation and complimentary close and begin directly.

When you are sensitive to your role—based on your relationship to your audience—you make organizational and stylistic choices that are appropriate to each rhetorical situation.

Purpose

Email can have a number of useful purposes. It can help you schedule meetings efficiently, find former college roommates, or disseminate copies of your most recent proposal. Generally speaking, the purposes associated with email messages include (but are not limited to) the following:

- To entertain
- To provide information
- To seek information
- To persuade
- To transmit attached documents
- To confirm plans

If you simply ask yourself, What do I want this email message to accomplish? and you provide a thoughtful answer, you'll have a clear purpose to guide the organization of your message. Recently, I wanted to know the origin of the phrase *I'm in a pickle,* so I emailed the Mt. Olive Pickle Company with my question. Because my purpose was to seek information from someone I didn't know, my message's organization was direct, my style somewhat formal, and my tone respectful.

Upon learning that I'd soon need new glasses, a friend who lives some distance from me sent the following email message:

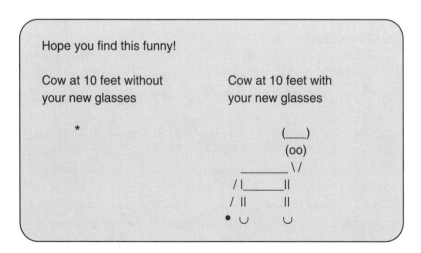

Clearly, my friend's purpose was to provide a chuckle, to entertain me. Because his style and tone were informal and casual, his organization was direct and his message was short.

Audience

Once you've established a clear purpose, you'll also need to consider your audience, or readers, and your relationship to them. Rhetoricians call this step in the communication triangle *audience analysis.* Whether you're communicating with your friends, your colleagues, or even strangers, taking a moment to create a mental profile of your readers and their preferences will guide you in determining the content and organization of your email message. Of course, if you know your readers well and the context is a personal rather than a professional one, most of your responses about audience analysis will be intuitive. Briefly, you need to consider these questions for your audience analysis:

- Who are your readers, and what are their preferences?
- What do they currently know, and what do they need to know?

Let's take a closer look at these questions. When determining who your readers are, you may want to consider their average educational level, experience, gender, age, socioeconomic status, and anything else that might help you characterize them. For example, the more education and experience your readers have, the more likely they are to understand complex concepts. The younger they are, the more likely you'll have to fill in their knowledge gaps.

Considering who you readers are allows you to formulate some ideas not only about the content of your email message but also about the format. For example, which of the following considerations do your readers prefer?

- Lists, headings, or highlighted text?
- Shorter messages or longer messages?
- The good news or the bad news first or last?

Related to this issue of preference, it can also help to ask, How will my readers use the information conveyed in my email message? Once again, your answer may be less structured if you are emailing information to friends or relatives, but if you are emailing business associates, teachers, government officials, or medical professionals, for example, you'll need to think carefully about your answer to this question. Most professionals appreciate receiving email messages that won't waste their time and will help them do their jobs better. Will your readers use your message to make decisions, perform duties, learn something about the organization, or fulfill job-related requirements?

Organizing an Email Message

In addition to considering your role, your readers, and your purpose, you'll also need to consider the most effective way to organize your message. Even though email is primarily an informal communication medium, specific conventions govern organization if your goal is for that message to be effective. More than 3,000 years ago, Greek rhetorician Aristotle referred to this part of rhetoric as *logos*. For Aristotle—and for today's writers—the logic of any message is the one component over which a writer has the most control. After all, you may not be able to change the way your readers perceive you. For better or worse, your status or your convictions reflect who you are and can color the way readers interpret your message. Likewise, your readers have their own biases, which you may or may not be able to change.

But you do have control over the way you organize the information in your electronic messages. You can choose a direct or indirect structure, depending on your purpose and your relationship to your readers. (As you can see, the rhetorical components of all messages are interdependent.) Let's examine these two structures—*direct* and *indirect*—as they relate to the most common types of email messages you'll write: good news, bad news, and persuasive email messages.

The most effective organization for a good news message follows a direct structure. That is, you want to convey the good news first and follow it with supporting details and a positive, forward-looking close. Compare the following two good news examples:

DIRECT GOOD NEWS MESSAGE

Dear Jo:
Congratulations! You've won $100 for your suggestion submitted to our "Save the City Money" contest.

The accounting department will issue checks on September 6, and you should receive your award a few days later in the mail.

We're excited by your suggestion. Thank you for participating!

Sincerely,

Chair, Suggestion Committee

INDIRECT GOOD NEWS MESSAGE

Dear Jo:
As you know, the city has solicited suggestions for its "Save the City Money" contest. It's been exciting to receive so many good ideas. Out of the hundreds of suggestions offered, we've chosen five.

I'm happy to report we adopted your suggestion. We will implement your suggestion at the start of the next fiscal year. You can be sure that your idea will make a difference to the city's budget.

To thank you for your participation in the contest, we will be sending you a check for $100. The accounting department will issue checks on September 6, and you should receive your award a few days later.

We're excited by your suggestion. Thank you for participating!

Sincerely,

Chair, Suggestion Committee

If the committee had selected *your* suggestion, what would you want to know first? If you're like most people, you'd want to know what you'd won and when you'd receive your prize. So what does that mean? Direct organization is usually the best choice for conveying good news.

But let's say you need to send bad news to someone. What's the best choice in that situation? To answer this question, think about a time *you* received a message conveying bad news—perhaps rejecting your job, school, or mortgage application. How did those messages start? Immediately with the bad news? Probably not. Most bad news messages usually cushion the blow with what's called a *buffer:* a short, neutral, non-confrontational statement that appears at the start of your bad news message. Buffers can show appreciation, as in "Thank you for your mortgage application." Buffers can also demonstrate fairness, as in "We have been reviewing the absentee rate for all employees." Writers who want to convey a sense of cooperation might construct a buffer such as "I'm sure we both agree the curriculum here is quite demanding."

Once the writer creates a buffer that's relevant to the topic, he or she follows it with the reasons and the bad news. Here's an example:

Thank you for your mortgage application [appreciation buffer]. Because your income does not meet the minimum requirements for the loan you seek [reasons], we are unable to approve your mortgage request [bad news].

Typically, after conveying the reasons and bad news, the writer closes the message with a positive, forward-looking statement, such as "If your income level changes, you are welcome to reapply. We look forward to serving your future banking needs."

If the writer had started the message directly with "We are unable to approve your mortgage request," the bank could have alienated the customer permanently. While the indirect approach cannot guarantee the customer will be happy to receive the news, the message's indirect organization is less likely to alienate the customer. Of course, there are times when the direct approach to bad news is preferable. For instance, if the message stays within an organization or if the writer and reader have a professional relationship, conveying bad news directly is acceptable. Direct bad news messages are more efficient because they are more succinct than indirect messages and thus take the reader less time to get to the point.

Persuasive messages can also follow direct or indirect patterns. In the direct pattern, the writer states the conclusion early and then follows it with the reasons. This approach is useful if you anticipate writing your email message to a receptive audience. When applying the indirect pattern to email messages, the writer first lists and explains the reasons before presenting the conclusion. This approach is effective when your audience is hostile or closed to your ideas. By first providing detailed information in support of your conclusion, you methodically prepare your audience to accept it.

Let's say you want to persuade your organization—your college, your place of work, or your parents—to replace some out-of-date computers. Because your place of work, for instance, may have a limited budget, you conclude that your readers will be closed to your suggestion. Knowing your readers' biases, you'll probably be more successful if you present your reasons before stating your conclusion that new computers are needed. Here's one way to organize this persuasive email message:

Dear Budget Department VP:

We both know how quickly technology evolves. As GMS consultants depend more and more on Internet access to remain competitive, it's important for us to access that information quickly.

While staying technologically up to date can be expensive at the outset, the increased efficiency of newer processors will no doubt increase productivity. In fact, just 6 months ago, MILDA Consulting spent $3.2 million dollars to upgrade employees' systems. And they've already doubled that investment due to increased contracts. That's not a bad investment. Given MILDA's success and our needs, I'd like to suggest that we upgrade all consultants' CPUs in the next two months.

Sincerely,

Marianne Reiher, Consultant

As you can see, the first paragraph sets up a general truth—that keeping technology up to date can be expensive. The second paragraph, however, provides two logically persuasive tools—figures and history—to argue that GMS can recover the initial expense. First, figures (specifically, the cost and financial return) support the argument about MILDA's success. And second, using history (MILDA as an example from the recent past) makes an argument in favor of GMS updating its computers. By providing these reasons, the writer has prepared the reader for the conclusion.

If you are emailing your parents from college and want them to update your computer, you might approach the same message directly rather than indirectly. If your parents are responsive to the legitimate needs associated with your education, you can consider them a receptive audience. In this case, your email message might take this organizational structure:

> Hope you're OK, Mom and Dad. Will you transfer $2,000 to my checking account? I need to update my computer. With all the work that's piled on, I need a faster computer, so I can access the Internet and my email account more quickly. What do you say?
>
> See ya!
>
> Andreas

In this letter home, Andreas presents his point immediately. After the greeting to his parents, he asks for the money directly.

So, what's the moral of this discussion about organization? If you consider the purpose for your email message and your reader's likely reaction to it, you'll be prepared to decide whether direct or indirect organization is more effective.

Formatting Conventions

Now that you know how to complete a situational analysis and effectively organize your email messages, let's talk a bit about what goes into a strategically formatted email message.

Headings

When you create a new email message, you'll see the following *headings*—"To," "Cc," "BCc," and "Subject"—which appear at the tops of new email messages. Figure 1.1 shows the template for a new message in the email software called Outlook.

On the "To" line, simply type your reader's email address. While email addresses are not case sensitive (meaning you don't need to pay attention to typing upper- and lowercase letters), you should be very careful to insert underlines and other symbols accurately.

The "Subject" line typically needs the most thought. Because numerous email messages crowd our electronic mail boxes every day, they compete with one another for attention. To this end, the subject line is your first defense against your reader's "Delete" key. Unless your position in an organization is important enough to guarantee your email messages will be read, create a subject that reveals the *specific focus* of each message in as few words as possible.

Sometimes, it helps to write the email message first and then go back and fill in the subject line. A vague subject line, such as "Basketball tickets," won't grab your reader's attention because you don't provide enough information. Starting with a noun and then adding a verb makes any subject line more specific. For example, you might start with the subject line "Basketball tickets." To make it more specific, add a verb; then "Basketball tickets" becomes "Basketball tickets *offered*." Adding appropriate modifiers can make a subject line even more specific. "Basketball tickets offered" becomes *"Free Ohio University-Ohio State* basketball tickets offered." If you're a fan of either team, you can see how much more enticing the specific subject line is compared to the first version.

Here's one more example of converting a general subject line to a specific one:

W-2 information → W-2 information provided → Important W-2 information provided

Another heading that requires your attention is the "Cc" line. Most email software programs allow you to send copies of messages to secondary recipients. The abbreviation "Cc" is a carryover from the "carbon copy" days. While "Ec," or "electronic copy," is not only more current but also more accurate, tradition prevails, so "Cc" is the heading you'll usually see in email software programs. If you want to forward an electronic

FIGURE 1.1   Template for a New Email Message

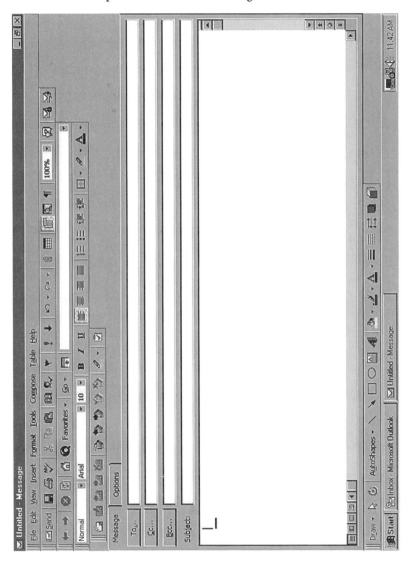

copy of your message to another reader, then you'll simply need to include that reader's address on the "Cc" line.

The "BCc," or "blind carbon copy," line allows you to send an electronic copy of your message to a secondary recipient without letting the primary recipient know. For example, let's say you're a middle manager. You want to send an email message to one of your employees, and you want your supervisor to know about it without letting your employee know you sent the second copy. To do so, you'd include your supervisor's email address in the "BCc" line. Your employee, the primary recipient, will be "blind" to the fact that you've forwarded the same message to a secondary recipient, your supervisor.

Consider another example: Suppose a middle manager in the human resources division of a major automaker is responsible for evaluating several employees semiannually. On one occasion, a furious employee challenges his evaluation, which is "below average" in four out of nine categories. Emailing the employee her justification for the evaluation, this middle manager also should send a "BCc" to the VP of human resources to document this potentially volatile situation.

While all the usual reasons for sending blind copies of paper correspondence continue to be valid for electronic mail, the "BCc" line has yet another function. If you want to be sure the hardware and software sent your message and that your recipient received your message, send a "BCc" to your own email address. When that message shows up in *your* "in" box, you'll know the transmission occurred successfully.

Salutations

Because email messages combine format conventions from both letters and memos, the issue of whether to use a salutation becomes interesting. You may remember that letters include salutations and memos do not. So, should you include a salutation in an email message?

The answer depends on your relationship to the recipient. If your message is going to friends, family, or colleagues whom you know well, either omit the salutation or simply use the person's first name. If you use the person's first name, you may write "Dear Kurt" or curtly write "Kurt." If your message is going to people whom you do not know well or do not address on a first-name basis, then include the salutation and appropriate title. If you're still unsure of what to do, adhere to standard formalities and simply create a salutation that accurately reflects the way you would address that individual in person.

Length

Keep your email message to one screen or one page, if at all possible. While information overload may seem like the primary reason for this guideline, there is also a psychological reason. When you sit down to read articles in magazines or chapters in books, if you're like most people, you flip to the end to determine the number of pages. It's human nature to do so. We want to see what we're in for. And knowing how long a particular piece of writing is tells us how long we'll have to concentrate to read it.

The same thing holds true for an email message. Before we read it, we want to know how long it is. But we can't physically handle an email message or quickly flip to the end of it. So, do your readers a favor: Let them see the end onscreen by keeping your email message short. If your message must be longer than one page, consider creating an executive summary or sending a "Heads up; it's coming" email message. Then forward a hard copy of the longer message by regular, overnight, or interdepartmental mail.

Also consider the length of each line in your email message, especially if you're sending it outside your business's, school's, or institution's network or if you're sending a message from home that may go through several different Internet service providers. Some terminals can't handle text longer than 80 characters per line. If your email software *word wraps* each line automatically—in other words, it breaks the text into consecutive lines—at a specific character limit that doesn't exceed 80 characters, then you'll be OK. Either count the characters in one of your typical email messages, or check with your technical support person. If your email software doesn't automatically word wrap at 80 characters, then you'll have to remember to hit the "Enter" key manually before reaching 80 characters. While this manual procedure may be annoying, it's necessary if you want your message to appear to your recipient the way it appears to you.

You'll also need to signal the end of your message clearly. This is important even when your reader can see the physical end of a short message. By ending with an obvious closing, you won't risk leaving your readers wondering whether you forgot to include something important or whether a snag in the transmission clipped off the message. If the message is long enough, signal the end by briefly reviewing its content. If your message requires specific action from your reader, succinctly remind him or her of a deadline, quickly reiterate a request, or concisely repeat a question. If the message is somewhat complicated, end by concisely reviewing its contents.

Once you've ended your message, add your name or initials. You also have the option to include a quick complimentary close, such as *Sincerely, Thanks, Cheers,* or *See you at lunch on Tuesday.* Any of these options will make it clear to your reader that no more information follows.

Signatures

Current technology won't allow you to sign your email in your own handwriting, which means your email signature is always typed. And given the hybrid nature of email messages—somewhere on the scale between a memo and a letter—much of the information about salutations applies to electronic signatures.

Typically, your relationship to the recipient of an email message will determine two things: (1) whether your message should contain a signature and (2) if it should, what form that signature should take. Keep in mind, however, that your position in relation to your reader changes continually if you send email messages on a regular basis. You may, for example, send messages to friends, colleagues, relatives, subordinates, and superiors—all in one day.

If you correspond frequently with someone whom you know well, it's OK to include only your first name or initials without the complimentary closing. However, if you want to maintain a formal or businesslike relationship with the recipient, it's a good idea to create a signature that reflects what you would include in a hard-copy signature. When your readers don't know you well or not at all, include a complimentary closing and your name along with your credentials. Knowing your title and position will likely cause your reader to take your communication more seriously.

Several other factors will help you decide whether to end your email message with a complimentary closing and formal signature with credentials. If your message originates from company equipment, your signature should include not only your email address and name but also your title and your organization's name. Doing so clearly signifies that you're using the organization's hardware and software for official communication. When it becomes common to send and receive email messages on letterhead, you'll only include the signature information not included on the letterhead.

Because some organizations use abbreviations, numbers, or a combination of both when they assign email addresses, the message you send might be identified cryptically in the "From" line once your message reaches its electronic destination. If you don't have an alias that clearly identifies you by name, your reader may not link the "From" line in your email message to you. Take a look at Figure 1.2 (page 40). You can clearly see the cryptic nature of the alphanumeric email address in the "From" line. Because the writer didn't include a name at the end of the message, it's difficult to know who sent it.

To be sure your recipient knows that *you* sent the message, make it a habit to include at least your name or initials in all email messages, regardless of your relationship to the recipient. Doing so avoids any possible confusion about who sent the message. Moreover, a signature clearly signifies the end of the message.

Also consider including your email address with your signature, especially if your email message is going to someone outside your organization. You certainly could argue that your email address is clearly indicated on the "From" line. You could also argue that the simple act of replying to a message eliminates your recipient's need for an explicit email address. After all, if someone emailed you the first message, then he or she presumably has your email address already.

But think of it this way: When you call someone and leave a voice mail message, you include your telephone number. If you're out of the office when that person calls you back, he or she leaves a message on your voice mail. Wouldn't life be easier if when you were ready to return the call, you were able to get the number from the voice mail message, instead of having to search for it? Surely, it would. So, do your readers a favor: Make email life easy. Including your email address with your signature ensures the address will be readily available, easily identifiable, and conveniently exchanged between email writers.

The problem is that it can take a lot of time to include a signature with your name, title, credentials, organizational affiliation, and email address in each message. To make this easier, your email software gives you the option of creating a *signature file,* which automatically inserts all this information. When creating a signature file, just keep in mind your relationship to your readers. It's best to create a signature file that works in all rhetorical situations—from casual to informal to formal.

More and more email writers are including quotations and characters—in addition to the items discussed above—in an attempt to

FIGURE 1.2   Cryptic Email Address

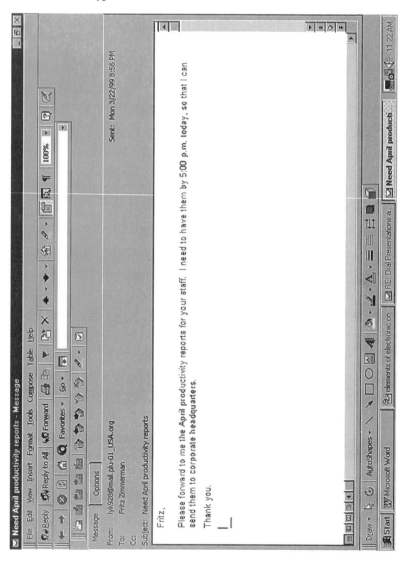

personalize their signature files. For example, a development director for a not-for-profit organization ends all her email messages with a quotation from Abraham Lincoln: "Determine that the thing can and shall be done, and then we shall find the way." I admire Abraham Lincoln; however, his ideas, as well as Indian proverbs and transcendental philosophies, are fascinating only in specific contexts—and email is not one of those contexts!

Because most email users suffer from information overload, include only relevant information in any email signature file. After all, it's in your best interest to keep your message to the point. Ask yourself the following question: Would you include quotations or characters in *hardcopy messages* written to business associates, teachers and professors, or potential clients? If the answer is no (and it likely is), then avoid doing so in email messages, too. If the answer is yes and you are compelled to include quotations and characters with your electronic signature, make sure they support the tone of your message. A serious message will be reduced by a light-hearted quotation, and a light-hearted message can't tolerate a serious quotation. Just be consistent in your tone, and you'll make the right choice when including quotations and characters with your signature.

```
           @ @ @ @ @ @
         @ @ @ @ @ @ @ @
          /  ~      ~  \
          {  (O)  ||  (O)  }
 ─────────────────── oOOOo ─── oOOOo ───────────────
```

One last point: If you include a quotation or a character with your signature, be sure to keep that information short and nonobtrusive. You want your reader to focus on your message, not on a long quotation or complex character.

```
 _____.oooO_____Oooo._____
                    (   )       (   )
                    \  (         )  /
                     |_)        (_|
```

Emphasis Techniques

Earlier, we talked about the increasing number of email messages most people receive each day. And we also talked about how quickly readers delete messages. If you create a message that enhances my ability to read it quickly and easily, that message is more likely to get my attention.

With this strategy in mind, let's look at some emphasis techniques that will protect your message from your reader's sweeping use of the "Delete" key. In general, observe the following suggestions:

- Present your main points quickly.
- Briefly preview the remaining content.
- Use bulleted lists, highlighted headings, and single-sentence paragraphs.

Presenting your main points quickly gets your information across to your reader more efficiently. Previewing the remaining content tells your reader what to expect. And offering information via bullets, highlighted text, and short paragraphs contributes to reading comprehension and efficiency. Compare the two examples of email messages in Figure 1.3 (pages 44–45). Of the two, which one were you able to read more quickly? Which one were you able to understand more easily? In the first example, the writer conveys the information in a single, traditional paragraph. While the information contained in that paragraph is clear, the information contained in the second example is much easier to digest because it is separated into bulleted items.

Style and Tone

Style and tone result, in part, from word choice, sentence length and patterns, figures of speech, and other linguistic choices, such as using all capital letters. The combined effect of these choices determines style and tone, which can be considered on a continuum that goes from formal to informal to casual. As with all writing, it's important to make intelligent choices about style and tone for email messages.

Let's consider an example: A senior economics major with a GPA just under 3.0 wrote an email message to an employment recruiter for one of the largest investment banks in the United States. The student's overly familiar tone made it sound as if he were chummy with the recipient of his email message (whom, by the way, he had never met). He called one of this recruiter's colleagues "a very sweet person," and he included several smiley faces in the message. The recruiter responded by writing, "Email correspondence should not be treated as a sticky note." Needless to say, the applicant didn't get the job. So, what's the moral of this story? Don't assume just because email is primarily an informal communication channel, style and tone don't matter. They do!

Email is a hybrid of written and spoken communication. As such, the conventions governing style and tone can be challenging. Should you adopt a casual style and tone appropriate to light-hearted circumstances? Should you adopt an informal style and tone appropriate to informal speaking situations? Or should you adopt a rigid style and tone appropriate to formal written contexts? The answer depends on your purpose for writing, your message, and your relationship to the audience who will read your message.

Having an appropriate style and tone clearly affect a message's effectiveness. For that reason, let's look at some general characteristics associated with each level of formality. Doing so will help you decide which style and tone are appropriate for your email messages.

Levels of Formality

How would you characterize the style and tone of the following email message sent to a company's stockholders?

> The impending liquidation of the company's assets will not take place until the end of next week, at which time the company's warehouse will be closed to everyone except registered participants. The company looks forward to a continued and prosperous relationship with its employees, its business partners, the community, and its investors and promises to devote resources that will help the company build up much of its remaining assets in the coming fiscal year.

Fairly formal, isn't it? A *formal* style and tone has these traits:

- Is serious
- Maintains a figurative distance between the writer and reader

FIGURE 1.3   Email Message without and with Bullets

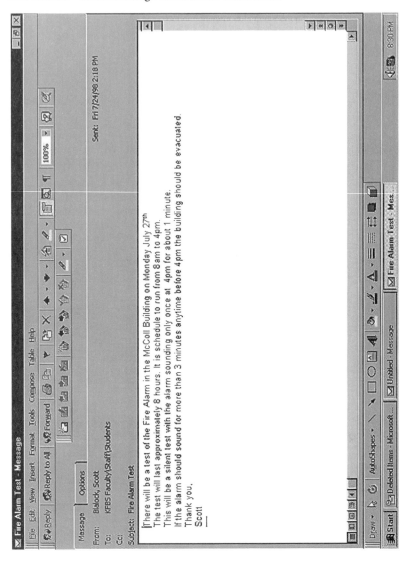

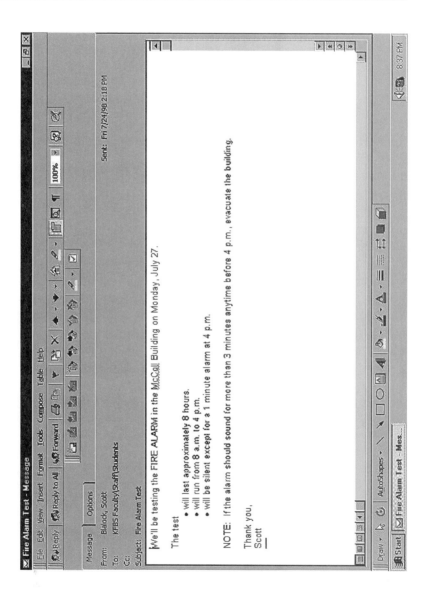

- Avoids personal references to maintain an impersonal tone
- Avoids contractions
- Incorporates abstract language in long sentences and paragraphs

Most of your email messages will avoid the stuffy, formal style and tone used in the preceding example, even when those messages contain attachments that require a high degree of formality.

The next level of formality can be labeled *informal.* Here's an example:

> Hi, Keith,
>
> Can you get the latest figures on the Performance case? I need them for our presentation at 2:00 p.m. Drop them on my desk. I'm heading out for a quick bite.
>
> Thanks, Martha

As you can see from the preceding message, an informal style and tone means that your email message has these qualities:

- Shows sentence variety
- Uses first- and second-person pronouns and contractions to achieve a more personal tone
- Demonstrates a slight figurative distance between the writer and reader

The style and tone of this book, for example, is informal. In fact, with a little effort, you'll be able to find all the characteristics on this very page associated with an informal style and tone. Keep in mind, too, that most of the email messages you'll write for professional and business purposes will adopt an informal style and tone. Doing so maintains an approach that is appropriate without sounding stuffy. A word of caution, though: Don't assume that using an informal style and tone gives you permission to ignore grammar, spelling, and word choice. It doesn't!

Next take at look at the *casual* style and tone of this email message:

> Hey, Bozo! What's happenin', dude? Matt and I are gonna hop down to Court Street and have us a brew. Wanna come? The libe will be there when we're done, man. We're leaving tonight at 11.

What do you notice about this email message that makes its style and tone casual? Clearly, it's quite different from the previous two messages, which demonstrated a formal and informal style and tone. If you choose to adopt a casual style and tone, you'll typically create a message that can be described as follows:

- Uses short- to medium-length sentences (20 words or fewer)
- Has short paragraphs (3 to 5 sentences)
- Incorporates descriptive language using specific words in conjunction with whimsical topics
- Uses numerous personal pronouns and contractions
- Refers to people by first names, pet names, and nicknames
- Incorporates slang, jargon, and colloquialisms comfortably
- Minimizes the distance between the writer and reader

When corresponding with close friends and relatives, your email messages can adopt a casual style and tone. However, don't allow casual writing to become obscene or otherwise indelicate. And be sure to reserve casual messages for those persons closest to you.

Flame Mail

One convention of electronic communication that contributes to an email message's tone is called *flaming*. People who send flame mail, or *flames*, often send their unpleasant email messages typed in all uppercase letters. While one or two words typed in capital letters is acceptable for conveying emphasis in a professional or personal email message, you should avoid messages written in all caps. Doing so is the electronic equivalent of an ear-splitting cussing out.

Flames can be direct, as in "YOU ARE A STUPID NITWIT," or indirect, as in "I THINK THAT'S A DUMB IDEA." In both examples, the flames are directed at a person rather than at an idea or concept. While it's acceptable to disagree with a concept or an idea via an email message, disagreeing with the source of that concept or idea—that is, the person who conveyed it—is inappropriate. Simply put, flames are unprofessional and impolite in all electronic communication contexts. So DON'T EVER, EVER SEND THEM! (Sorry I yelled.)

But, you might ask, what if I read an email message that angers, irritates, or otherwise upsets me? Whatever you do, *don't respond immediately.* You might be tempted to flame someone. Instead, wait 24 hours to regain your composure. Then formulate a professional response, as if you were explaining politely and respectfully to your Great Aunt Elly (for the 103rd time) why the mold on the cheese she tried to serve you is unappetizing. Just remember: Don't hide behind your computer terminal with your "Caps Lock" button pushed.

Sometimes, you might be blamed for sending a flame even if you didn't intend to. If this happens, simply apologize by explaining that you didn't mean to convey that impression and you're sorry the message was misinterpreted.

Now you know how to avoid flaming someone else. But what techniques should you observe to avoid being flamed? First, don't offer negative comments about someone else's email message. Many people still have the erroneous notion that email tolerates a relaxed communication style and tone, in which attention to correctness is not valued. As a result, email messages often contain more grammatical and mechanical mistakes than found in other more formal types of communication. Even so, never, ever express a view about someone else's grammar, mechanics, word choice, or writing style if you want to avoid getting a flame message in return.

Another way to avoid being flamed is by not using the "*ALL" (or "star all") function carelessly. Limit its use to *responsible* mass mailings. You can expect to get flamed if, for example, you offer your opinion on recent political developments to all the seniors at your high school, ask folks at the office to buy your daughter's Girl Scout cookies, or attempt to sell cosmetics or cookware or anything else via the hospital network. When in doubt, follow the "golden rule" of email: Email unto others only what you would have them email unto you!

Acronyms

Email has developed its own language filled with esoteric acronyms and cryptic symbols most likely developed to make email convenient and efficient. Using acronyms allows us to type information quickly, IMHO. However, if acronyms obstruct rather than convey meaning to readers, then how truly efficient are they? Wouldn't you be more apt to understand "In My Humble Opinion" rather than "IMHO"? Certainly, incorporating acronyms in your messages will save you some time during the writing process, but it will cost your readers much time if they have to figure out what you mean. Since the purpose of any message is to get results, be as clear as you can.

So, how do you decide whether to include acronyms? The answer is simple. If the acronym you want to use is not standard in everyday English, avoid it. Besides, if you keep your email messages succinct, then acronyms will be unnecessary. However, technical acronyms are useful for email communication with systems administrators. Table 1.1 lists

TABLE 1.1   Technical Acronyms

Acronym	Full Term
ASCII	American Standard Code for Information Interchange
BBS	Bulletin Board Service or System
FAQ	Frequently Asked Question
FTP	File Transfer Protocol
GIF	Graphics Interchange Format
HTML	Hypertext Markup Language
HTTP	Hypertext Transfer Protocol
IRC	Internet Relay Chat
ISP	Internet Service Provider
JPEG	Joint Photographic Experts Group
URL	Uniform Resource Locator
WWW	World Wide Web

Note: See the Guide to Specialized Terms for Electronic Communication (pages 9–18) for definitions of these terms.

some of the technical acronyms that are becoming common among email users. The point to stress here is *know your audience.* If you do, you'll use acronyms appropriately.

Emoticons

When you communicate face to face, your words convey only a portion of your message. Your body language—including posture, facial expressions, eye contact, and hand gestures—reveals whether you're interested or bored. And your voice—including articulation, inflection, pace, and volume—reveals whether you're enthusiastic or secretive. A slow pace, for example, communicates deliberate emphasis, while a fast pace communicates excitement about your message. Taken together, the way you look and the way you sound give emotional significance to your words.

As is true of any written communication, an email message cannot convey the meaning possible with a face-to-face conversation. Rather, in written communication, the meaning of your message depends exclusively on the words you've chosen. In an attempt to attach emotion to electronically communicated messages, email has adopted symbols called *emoticons.*

The word *emoticon* was created from *emotion* and *icon.* Writers construct emoticons from keyboard characters, which are carefully placed after the specific words, sentences, or paragraphs to which the emoticons refer. The emoticons attempt to replicate what your readers hear and see when they communicate with you face to face.

For emoticons to make sense, you have to look at them sideways. For example, a colon followed by a hyphen and a closed parentheses creates an emoticon that conveys delight. A colon followed by a hyphen and an open parentheses creates an emoticon that conveys sadness. Some word-processing program can even convert these symbols into actual smiley and frowny faces:

:-) = ☺
:-(= ☹

Now that you know about emoticons, should you use them in your email messages? The answer depends on your purpose for writing, the subject of your message, and your relationship to your readers. If you are writing within a professional, formal, or business context, avoid using emoticons. They add a level of informality that may trivialize any email message. If you are writing within a personal, informal, or playful context, however, emoticons can be acceptable if you don't overuse them.

Just a few more words of caution: Too many emoticons can be distracting, intrusive, and annoyingly silly. Moreover, using emoticons can be risky. Even though the purpose of using them is to capture your emotion, your reader may interpret them differently. If you decide to use them, even in casual messages, be sure the emoticons hold the same meanings for your readers as they do for you.

And one more piece of advice: Never, ever substitute an emoticon for a carefully chosen word or phrase. Not every picture—certainly not a rudimentary picture constructed from keyboard characters—is worth a thousand words! In appropriate situations, if you stick to the emoticons in Table 1.2, you'll probably be OK in terms of successfully conveying your intended meaning.

Punctuation

As with emoticons, using too many unnecessary punctuation marks can be intrusive in an email message. The key word here is *unnecessary*. I'm not suggesting that you eliminate necessary periods, semicolons, and commas. Rather, I'm suggesting that you not overuse exclamation points (known as *bangs* in computer circles) and question marks for emphasis. Again, your relationship to your audience, your subject, and your intended purpose for writing the email message will determine how to use punctuation marks.

TABLE 1.2   Common Emoticons

Emoticon	Name	Meaning
:-) or :)	Smiley	Expresses pleasure and delight, softens criticism, or hints at sarcasm
:-(or :(	Frowny	Expresses sadness or anger or hints at empathy or sympathy
;-)	Winky	Expresses "You know what I mean" or indicates "Just joking"
:-0	Shocked	Expresses shock or surprise

If you want to emphasize something, let your word choice, not your punctuation, communicate that meaning. If you are so anxious to get the answer to an email question that you end the question with 20 question marks, you'd be better served to pick up the telephone and ask the question directly. If you are so hysterically agitated or excited that you need 20 exclamation marks to exorcise that emotion, you'd be better off to go to the top of the nearest uninhabited mountain and scream. Don't get into the habit of relying on punctuation marks to convey excessive inquisitiveness, impatience, or anxiety. Doing so will make it harder for you to shift from writing casual email messages addressed to close friends to writing email messages addressed to professional associates, teachers, and business contacts.

Responding to Email Messages

We've spent a good deal of time talking about creating email messages. Now let's discuss the proper way to respond to the email messages that show up in your box. Because your response needs to be connected in some way to the original message, never create a new message if you're responding to an earlier email message. Many of us tend to get multiple messages from the same person in a given day. So when you respond, it's important to help your reader know which message you're referring to.

Think of it this way: If you receive three questions from the same person via three separate telephone calls and each time you tell the caller you'll respond to the questions later that day, what will happen when you call back later and respond to question two without referring to that question? Your listener will be temporarily confused. An effective response, in any rhetorical situation, is sensitive to the entire communication context associated with the particular message. That always includes a reference to the original message and to the response.

Most email software makes this requirement easy by automatically including a copy of the entire original message as part of a response. If you click on the "Reply" or "Respond" button in your email program—which appears when you've opened any email message sent to your "in" box—your response will appear just in front of the original message. In other words, when you hit the "Send" button, your response and the entire original message will *both* go back to your reader, in that order.

When responding, sometimes it's better simply to quote the relevant parts of the original message and not the entire message. If, for example, the original message is long or includes several issues, quote only the single issue from the original message related to your response. Doing so helps the recipient put your response in a clearly focused context and allows your reply message to be more succinct.

Before we leave this section on responding to email messages, I have one last comment: Not everyone or every organization can respond to email in a timely manner. While businesses around the world accept email from consumers, they have a hard time responding to those messages once they arrive. It's not just that email messages are misrouted, lost, or responded to incorrectly. It's that the sheer volume of messages is overwhelming for many organizations. Consider a few examples:

- About a month ago, the German air carrier Lufthansa advertised that it would hold an auction for international airline tickets in late February. Because I wanted more information about the process, I emailed the company. By mid June, I had yet to hear from Lufthansa. I assume the auction happened without me.

- Earlier I told you that I emailed Mt. Olive Pickle Company because I wanted to know the origin of the phrase "I'm in a pickle." As with Lufthansa, I have yet to receive an email from Mt. Olive.

- President Clinton receives 2,500 email messages a day. Do you know how many responses are sent out in his name? Zero.

What's the moral of the story? Email may turn out to be less efficient than we thought. What's the other moral of the story? Don't neglect using the telephone or U.S. mail.

Issues Regarding Email

As we draw our discussion about effective email messages to a close, we should take a look at a few more issues: a pervasive email myth, mass mailings, and privacy, legal, and ethical issues.

Can Email Ruin Your Computer?

The Internet is full of warnings about viruses that claim they'll infect your computer, destroy your hard disk, and physically damage your computer's parts. But the majority of these warnings, usually transmitted to your email "in" boxes, are usually nothing more than pranks, designed by bored and malevolent individuals to create panic in cyber-space. Just because these warnings come from the Internet doesn't mean they are legitimate.

Because these messages seek to scare the largest number of computer users possible, they usually come with urgent pleas—punctuated with excessive exclamation marks and written in all uppercase letters—to forward the message to everyone you know. For instance:

> DO NOT OPEN ANY MESSAGE TITLED "GOOD TIMES" OR YOUR HARD DISK WILL BE WIPED OUT!!!!! TELL EVERYONE YOU KNOW!!!!!

This extreme level of urgency is another indication that the warning is bogus.

To aid you in determining the validity of these warnings, Rob Rosenberger, web master for the Computer Virus Myths home page, suggests applying three common-sense tests in evaluating any virus warning:

1. Ask yourself who wrote the message.
2. Determine for yourself whether the message makes sense.
3. See whether the message urges you to distribute it to all your friends.

If the message is anonymous or from a stranger, not from someone whom you know or know of, you can probably discount it. Any legitimate warning will come from your system administrator or Internet service provider (ISP). Be aware that many email hoaxes about computer viruses will attribute the truth of the virus and its warning to authority figures. They may say something like "Norton Antivirus technicians first mentioned this virus earlier this month" or "Officials at Microsoft announced the virus warning at a recent press conference." Often these authority figures have made no such announcements, are quoted without their consent, and may not even know that their names, positions, and corporations are being used for these purposes.

Many false warnings are also often written in bureaucratic, official-sounding language that really makes no sense at all. Remember, it's OK to read the message because doing so cannot destroy your computer. However, if you don't understand the warning once you read it, don't blame yourself for computer illiteracy. Rather, question the message. Any authentic warning will be written in a way that makes sense and will come from a reputable, verifiable source.

There is, however, a legitimate concern regarding email and viruses. If you receive an email message with an attachment and that attachment contains an infected software file, opening the file may infect your system with a real computer virus. One such example is the "Melissa" computer virus, which in March 1999 infected more than 100,000 computers in hundreds of companies—including Du Pont, Lockheed Martin, Honeywell, and Compaq Computers. The email message through which the virus was spread contained the subject line "Important Message from [Someone you probably know]" and a 40K Microsoft Word document attachment called "LIST.DOC." Opening that attachment infected Word, scanned your email address book for the last 50 people you sent email to, and then sent each of them an email, which appeared to have been generated by you. The virus also infected the NORMAL.DOT template in Word.

Keep in mind that you have to open the attachment to release the virus. Look at Figure 1.4 (page 56), which demonstrates how an attachment can infect your software. Fortunately, many email programs come with built-in warnings that appear when you want to open an email attachment, as in Figure 1.5 (page 57). By saving the file to a disk, you protect yourself from viruses that may be sent inadvertently in attachments.

Email may eventually evolve into what virus expert Rob Rosenberger describes as a *live document* that absorbs everything, including attachments. Then warnings like "It's the Good Times virus for real!" will be, well, real. Until that time, Rosenberger reassures us, "You can't get a computer virus just by reading an email with your eyeballs."

Spam

Spam is the electronic equivalent of those pesky unsolicited mass mailings—copious offers for term life insurance, infinite credit card applications, and pervasive gadget and lingerie catalogues—that crowd your mailbox. It's similar to the junk mail delivered by the U.S. Postal Service. Spam got its name from the skit performed by the British comedy troupe Monty Python, in which the word *spam* was interjected intrusively and annoyingly into a discussion about breakfast foods.

FIGURE 1.4   Viruses from Email Attachments

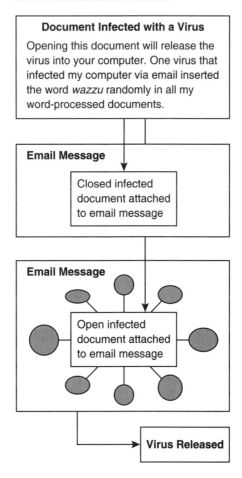

Intrusive and annoying, electronic spam floods the Internet and your electronic mailbox with countless messages. It's annoying because it forces you to deal with uninvited solicitations written primarily to sell things. Unsolicited commercial electronic mail, or spam, is easy to spot because most messages share the following characteristics:

- Are seedy and promote shady products, bizarre miracles cures, quasi-legal services, get-rich-quick schemes, or kinky pornography

FIGURE 1.5   Opening an Email Attachment

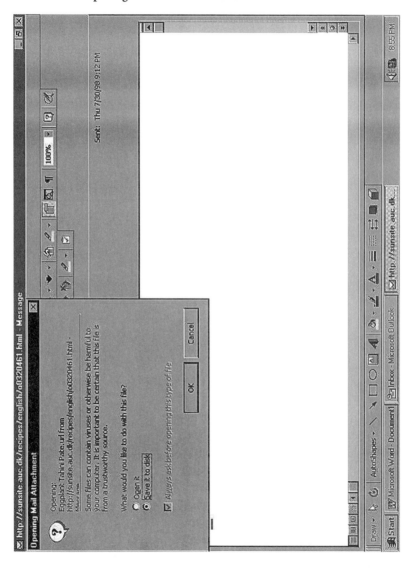

- Come to you without your consent
- Convey a sense of urgency through excessive use of exclamation marks!!!!!

Spam finds its way into your email inbox in one of two ways: as email spam or as Usenet spam. Email spam finds you directly by getting your email address from newsgroup postings, from Internet mailing lists, or from the Web. Usenet spam sends a single message to 20 or more newsgroups. Because a newsgroup can have many subscribers, some of whom post and some of whom lurk, the pervasiveness of a single spam message becomes evident. Indeed, when that pervasiveness is multiplied by hundreds of spam messages—sent either directly to you or through the help of a newsgroup—the problem becomes clear. Too many irrelevant messages can undercut email's purpose of direct and convenient communication, forcing you to spend valuable time deleting irrelevant email messages.

In addition to the reduced efficiency caused by span, it's harmful because of the costs associated with it. If you have measured phone service and read or receive your email while the "phone meter" is running, you, and not the spam sender, pay for the privilege. Moreover, transmitting spam messages costs your ISP money, which your provider then passes on to you in the form of increased charges for monthly access.

Fortunately, Congress enacted a bill called the Unsolicited Commercial Electronic Mail Choice Act of 1997 to help protect you from spam. Because the frequency of unsolicited email advertisements (aka spam) has grown exponentially, Congress has mandated that the word *Advertisement* appear in all unsolicited email messages. Spammers who don't comply with the law are subject to fines of up to $11,000. In addition, the name, physical address, electronic mail address, and telephone number of the person who initiates the commercial message transmission must appear in the message. This requirement holds spammers more accountable for their messages. It also allows you to delete unwanted messages more efficiently. By simply reading the "Subject" line, your electronic filter can block spam.

While ISPs are not subject to this law when they forward spam, they are subject to the law if they initiate spam transmissions. Keep in mind, however, that if you respond to an unsolicited email message for goods or services, you've authorized its transmission, so the spammer isn't breaking the law in that situation.

One final word of caution regarding spam: It's certainly legal for you to request, by electronic mail, that spammers stop sending you unsolicited email transmissions. And the spammer who receives such a message has 48 hours to stop. However, be careful about making this electronic request. Doing so may reconfirm your email address for the spammer, who may then increase, rather than eliminate, unsolicited mailings. Instead, you should email or call your ISP to register your complaint.

Privacy, Legal, and Ethical Issues

In the late 1970s, as a result of the Watergate scandal, the federal government created the Electronic Communication Privacy Act (ECPA) to control excessive eavesdropping on telephone conversations. Since that time, the ECPA has evolved to address privacy issues related to electronic communication. Today, the ECPA prohibits unauthorized interception or disclosure of email by *all* persons, businesses, and government not parties to the communication.

The first reported case concerning email and privacy rights came about in 1996 in Pennsylvania and involved Michael Smyth, who worked for Pillsbury. Pillsbury's company policy assured employees that email messages were confidential and the company could not use them as grounds for termination. However, when Smyth sent a biting message to his supervisor, characterizing other colleagues as "backstabbing bastards," the company fired him. Smyth sued Pillsbury in federal court, alleging wrongful discharge, but he lost the case. Under Pennsylvania law, the court said as that soon as Smyth mouthed off inappropriately in cyberspace, he lost any "reasonable expectation" for privacy. Smyth's case against Pillsbury illustrates what most lawyers currently believe about email messages written on company systems: Unless your company explicitly gives you the right to privacy on the company email system in *all* written contexts, you don't have that right.

The ECPA's protection can be somewhat nebulous because it contains exceptions. If a system administrator has reason to believe illegal activity is occurring over the system, he or she may disclose the information to legal authorities. Computer programmer Augustus Lindsey from Oklahoma learned this lesson the hard way when he sent the following message to a colleague: "That little sex kitten has been driving me wild. She's moaning and begging for it every minute. Last night I was afraid someone would hear, and we'd be thrown out of the building. But don't worry—all is arranged. Wednesday she gets the knife." Thinking a crime was about to occur, Lindsey's supervisor alerted the police, who arrested the computer programmer. Police released Lindsey the next morning, just in time for his veterinarian to spay his cat.

Despite the good intentions of the ECPA, email is not private at home or at work. As I mentioned earlier in this chapter, some ISPs keep copies of your email messages long after you've deleted them. And more and more businesses are keeping employee email messages forever! It's becoming clear that computers, like elephants, never forget.

According to an American Management Association survey, 15% of employers store and regularly inspect company email messages. But the actual number of U.S. businesses in industries such as banking, insurance, and telecommunications that eavesdrop on employees' email messages is estimated to be higher than 25 percent. These companies want to monitor "cyber-lollygagging" to ensure employees aren't wasting valuable work time sending and receiving trivial messages. Moreover, companies are concerned with internal security and want to be sure employees aren't leaking company secrets. So systems administrators have the authority to intercept and monitor not only the flow but also the content of email messages, often without employees' knowledge or consent.

How can you protect yourself? You can start by finding out the email policies of your ISP and your employer. American Airlines, Federal Express, and United Parcel Service, for example, have email systems that automatically tell employees the company reserves the right to monitor their email. But today, corporations with such privacy policies are in the minority. In fact, only about one-third of companies currently have email policies, even though the Electronic Messaging Association has, since the early 1990s, been urging organizations to make policies readily available to all employees.

To help you clarify appropriate email use—whether at school, at work, or even at home—here are some questions to ask your email system administrator:

- Is email use exclusively for business, educational, or government use? Or are occasional personal email messages acceptable?
- Does the company reserve the right to review, audit, and disclose my email messages?
- What kinds of messages are considered derogatory, defamatory, obscene, or inappropriate?

- Does my deleting a message or file completely eliminate it from the system? If not, how long does the organization maintain the message?

- What monitoring policies does the company employ?

Even with answers to these questions, there is, of course, another issue: Regardless of the law, is it *ethical* for a system administrator to read someone's mail? Proponents say if the organization owns, operates, and repairs the equipment, then it's justified in reading employees' mail to ensure they are using the technology appropriately. Others argue that reading someone else's mail is simply wrong, and no organization is on sound ethical ground just because it tells people it may monitor their email messages. No matter where you see yourself on this ethical continuum, remember that *email is not private*. And even if it were, no email message should contain anything that might cause your readers to feel threatened, discriminated against, or sexually harassed.

It's also important to remember that email messages can sometimes go to the wrong audience. For example, while it is illegal to do so, hackers can intercept, read, and forward your email messages. Not only that, email software—even YOU—can malfunction, causing your message to go to an unintended reader. In one embarrassingly funny example, recently published by *The Phoenix Gazette,* an email user sent the following message to thousands of online computer users rather than to his girlfriend: "I would love to kiss your hose-covered toes." While that message might make a hosiery company happy, it caused the boyfriend to run, high-tailing it out of the office for the rest of the day!

Certainly, organizations that transmit sensitive or confidential information—such as personnel actions, performance appraisals, and patient information—rely on encryption software to ensure the information they transmit via email remains private. While encryption software can help protect the privacy of email messages in the workplace, you should check with your organization's technology person before installing it on company hardware for your personal messages.

If you're worried about privacy on your personal computer at home, encryption software may protect you. Still, if you simply think of all your email messages as public documents, you'll protect yourself from possible embarrassment, litigation, or discipline. Just remember: Never write anything in an email message that you wouldn't write on a billboard. Email is not private.

Summary

Benefits

- Email is convenient, fast, and inexpensive.

- Email is effective when you don't need to speak to your reader(s) face to face, when the recipient of your message isn't immediately available, or when the information you're sending isn't sensitive or personal.

Limitations

- Email cannot convey facial expressions, gestures, or speech patterns.

- Email cannot transmit your actual signature.

- Email isn't private. Familiarize yourself with privacy policies when using an organization's equipment.

- Because email isn't private, don't send messages when the information is personal, confidential, or inflammatory.

"Nuts and Bolts"

- Before writing an email message, determine your purpose and your relationship to your readers.

- Email addresses should be professional.

- Create a specific "Subject" line using a noun, verb, and modifiers.

- Salutations and signatures should reflect your relationship to your audience.

- Organize your message by presenting your main points quickly and previewing the remaining content.

- Keep your email messages short.

- Include bulleted lists, highlighted headings, and single-sentence paragraphs.

- Use emoticons only in casual email messages.

- Choose an informal style and tone for most messages.

- Don't send flames or spam.

- Use only those acronyms found in Standard English.

- Quote the original message or the appropriate part of it when responding.

2

The World Wide Web

The *World Wide Web (WWW),* or simply *the web,* is the fastest-growing component of the Internet. With special software and a connection to the Internet, you can create and distribute web sites that promote your products, services, and ideas to potential customers, scholars, and information-seeking individuals. What's particularly effective, though, is your ability to create a multimedia site that distributes not only text but also graphics, audio, and video.

Even the most technically advanced web site, however—loaded down with multimedia effects—will not effectively communicate to its audience if the *text* is poorly written, carelessly organized, or thoughtlessly laid out. Because Internet users have hundreds of thousands of web sites from which to choose, they have little patience with sites they cannot easily understand or navigate. According to Jakob Nielsen and John Morkes, two web scholars involved in web usability studies since 1994, web readers prefer succinct web pages enhanced by solid, easy-to-understand writing. So, in this chapter, we'll focus on how to create effectively written sites for the World Wide Web, and we'll also examine how careful organization and design contribute to a clearly communicated message. We won't, however, spend time on the technical skills required to create a web page. That kind of information is readily available online, from technical support staff, and through most software packages.

Purpose

The first step toward having well-written text on your web site is to determine your site's purpose. It's fine to want a web site because everybody else has one or because you're drawn to the technology. But you'll find it more productive to establish a specific purpose that guides all the choices you'll have to make if you want your site to succeed. Do you want to inform your readers, entertain them, persuade them to take some action, convince them to buy a product, or perhaps do a combination of these things? The more specific you can be about your purpose, the better prepared you'll be to create a web site that's rhetorically effective.

An effective web site can have several purposes for the people who visit. It can help you find an estranged family member or former college roommate, it can distribute information about your product or service, and it can argue—through text, graphics, video, and audio—for a specific political or social conviction. As with email messages, the purposes associated with web sites include the following:

- To entertain
- To provide information
- To persuade

If you simply ask yourself, What do I want this web site to accomplish? and you provide a thoughtful answer, you'll have a clear purpose to guide the organization, content, style, and tone of your site. To help you begin to focus your purpose, you may find it useful to complete the statement in Figure 2.1.

Commercial web sites—designed to sell products—often combine persuasion with entertainment. While Procter & Gamble's Tide Clothes-

---

FIGURE 2.1   **Possible Purpose(s) for Your Website**

The purpose of my web site is to _____.

(Pick the relevant purpose)

　　　　To entertain　　　To provide information　　　To persuade

Line site promotes detergent, it also offers a laundry contest and laundry trivia. Even the site maintained by the Internal Revenue Service (IRS) makes an attempt to entertain—reminding visitors the number of days left until April 15 and claiming that the site is "FASTER THAN A SPEEDING 1040-EZ"—while it provides important tax information.

Other merchants combine information about their products with persuasive sales pitches. Dell Computer Corporation invites visitors to explore the company's products, build a customized system, and pay for purchases. Similarly, Saturn offers visitors the opportunity to "Build Your Own Saturn," apply for credit, and calculate monthly payments.

Persuasive web sites, however, can do more than sell products or services. They can also persuade visitors to take up a cause or fund a campaign. Both sides in a heated debate, for example, will often post sites on the web to argue their respective positions. For example, the National Rifle Association's (NRA) site indicates that its members "fight hard to keep special interest groups from infringing on our Second Amendment rights." The site maintained by the Anti-Gun Coalition of America (AGCA) presents the other side of the issue, focusing on gun control, antiviolence, and antigun legislation. The first page of AGCA's site provides "a list of people, companies, and politicians who[m] the NRA has labeled enemies. Any enemy of the NRA is a friend of the AGCA!"

Audience

Once you've established the specific purpose for your web site, you'll need to decide whom you want your site to reach. Some, such as Wal-Mart's site, communicate with the public at large. Many newspapers and radio and television stations also have established web sites to disseminate news, weather, advertising, politics, business, and other information to the general public. Sites maintained by the *Washington Post,* National Public Radio, and CNN, for example, are an important part of those organizations' efforts to serve newspaper readers, radio listeners, and television viewers. Government agencies also maintain web sites to communicate important information to citizens. The U.S. Department of Education, for example, maintains a web site that offers financial aid information to prospective college students. The site maintained by the U.S. Department of Housing and Urban Development provides advice for buying a home and includes an exhaustive listing of homes for sale around the country.

On the other hand, some web sites are designed to reach specialized audiences. For example, the Amateur Entomologists Society, dedicated to the study of bugs, publishes a web site called *The Bug Club*. There, interested visitors learn how to care for caterpillars, cockroaches, and stick insects. Fan clubs maintain web sites targeted only at people who share their affection for, let's say, Al Pacino or Marilyn Monroe. Another web site may be intended for the small group of people who live in a certain neighborhood or belong to a certain income bracket. For example, the Cabrini-Green public housing complex, built 40 years ago on the near north side of Chicago, maintains a web site that publicizes the Cabrini-Green Tutoring Program, attended exclusively by students who live in or around the complex. Rolls Royce, at the other end of the income scale, targets wealthy computer users, clearly establishing its target audience with text that describes the cars as "exhilarating, high performance, [and] gloriously romantic."

Still other sites are so exclusive that they are available only to people who know their special passwords. For instance, government emergency management officials in coastal states maintain a private web site to exchange information on hurricane preparedness that is not accessible to the public.

By carefully analyzing your relationship with your target audience, you can determine many characteristics about them. You may determine that your audience consists of friends, peers, or professional colleagues who share interests or experiences with you. As such, your site will reflect a narrow focus. If, on the other hand, you are developing a site with a broad topic, such as the *American family*, you'll probably be targeting an audience from all social, religious, and economic backgrounds.

You may also decide to design your web site for multiple audiences, again depending on your purpose. For example, Bank of America's web site has sections reserved for customers who want to pay bills and check their account balances, sections that provide financial advice for small business owners, and still other sections with public relations contacts and press releases for reporters. BellSouth, a credit card and telecommunications company, communicates with multiple audiences. It allows prospective customers to order phone service online and allows current customers to review their phone bills—all online. Online travel agents, such as Cheap Tickets, introduce first-time visitors to weekly specials but require those who want specific information on fares to register. Even colleges and universities use the web, not only to help students register and pay for classes but also to market themselves to prospective students. As you can see, a web site can target multiple audiences.

Indeed, it's important to have some idea of your intended audience or audiences before you begin constructing your web site. The answer to Who is your audience? will, at times, be intuitive—for example, when you're writing sites for friends and family. At other times, however, the answer will require extensive research and costly analysis if you don't know. Answering the following questions—as honestly and thoroughly as you can—will give you a jump-start for your audience analysis and help you create an audience profile:

- What does your audience know?

- What does your audience need to know?

- What is the gender breakdown of your audience?

- What is the range and average age of your audience?

- What is the range and average income of your audience?

- What geographic region are you targeting?

- What are the biases of your audience?

- What motivates your audience to act certain ways?

- What software and hardware capabilities is your audience likely to have?

Certainly, additional questions may be relevant to your specific situation. Even so, I want to repeat: Your thoughtful answers to these questions will help you create and maintain an effective web site.

Unfortunately, knowing your audience isn't enough. Your site also needs to be *responsive to your audience's needs.* You can accomplish this goal in several ways. Dell Computer Corporation offers an online user survey designed to "improve your experience at Dell.com." Sears offers a feedback form for user comments on its site. Still other sites invite users to offer suggestions through "Tell us," "Contact us," and "Email us." In addition, if you offer your visitors something in exchange for filling out a short questionnaire on your site, you'll probably get information from more of them than if you don't.

By allowing your audience to contact you through your web site, you accomplish at least two goals:

1. You will have some indication of how visitors use your site. Two-way communication offers you an invaluable opportunity not only to respond to your audience but also to achieve bigger and better results by changing your site to accommodate audience needs and expectations.

2. Two-way communication allows visitors to inform you of problems with your site's construction. Based on this important feedback, you can update, revise, and reorganize your site to fit the needs, expectations, and interests of your audience.

So consider including a feedback section on each page of your web site. At the very least, if your site represents an organization, include contact information, such as the contact person's name and the organization's address and telephone number.

Style and Tone

Before you begin to design and write the text for your site, you'll need to make some choices about your site's style and tone. *Style* refers to the unique way you express yourself and is influenced by word choice, sentence structure, and paragraph organization. *Tone,* on the other hand, refers to the way your audience hears you. Rhetoricians agree that conveying information in a positive tone—even when the information you're conveying is negative—is more effective than the opposite. That is, rather than tell your visitors what you and your site *can't* do, focus on what you and your site *can* do.

For example, let's assume that you're selling beautifully handcrafted glass paperweights through your site and that I'd like to receive one before October 12. Let's also assume that you are temporarily out of stock. As you revise your site's home page, you consider including one of the following statements:

SENTENCE 1

Due to an increased demand for our beautiful handcrafted glass paperweights, we will be unable to ship new orders until January 20.

SENTENCE 2

Due to an increased demand for our beautiful handcrafted glass paperweights, we will be able to ship new orders beginning January 20.

While I might not like the news conveyed in either version, Sentence 2 conveys to me what you will do, which has a more positive impact than does Sentence 1.

In addition to conveying a positive tone, your site must also not oversell itself. That is, your tone must come across as sincere. For example, adopting a promotional tone, even for a commercial site, will reduce your site's credibility. Visitors can see through phrases such as *best ever* and *one-time opportunity*. While they hate exaggeration, visitors want information quickly conveyed through objective language. If they'll have to filter out the hype, they'll have to work too hard to get at the heart of your message. Too much filtering will cause mental overload and will turn off your visitors, who will be more apt to turn off your site.

Purpose and Audience

To a great extent, both style and tone are driven by the choices you make regarding purpose and audience. If, for example, you want your site to entertain your peers (Purpose = To entertain; Audience = Peers), then choose a casual, chatty style and a humorous tone. If, on the other hand, you want your site to persuade the public at large (Purpose = To persuade; Audience = General), then your style should be elevated and formal and your tone serious.

The Federal Bureau of Investigation (FBI) maintains a web site that lists the nation's 10 most-wanted fugitives, including their photographs and detailed descriptions of their crimes. As you might expect, the site has a serious style and an authoritative tone. Sites for popular comic strips, however, convey a whimsical style and humorous tone. At the Peanuts site, visitors not only get a look at the latest strip, but they also get a listing of upcoming "Snoopy Sightings." Sites that target children also convey a capricious style and light-hearted tone. The site maintained by the ice cream company that produces Bomb Pops tells young visitors if their tongues aren't pink, they "are not enjoying the taste sensation known as The Original Bomb Pop," which "paints your mouth with a rainbow of colors." Brokerage firms, on the other hand, adopt a formal style with an educated tone to inspire confidence in web site visitors. The site maintained by Charles Schwab tells visitors the site provides "the service you expect, the value you want" by guiding your investment selection through industry analyses, stock research, and market reports. Sites that target so-called Generation X-ers adopt an informal style with a hip, laid-back tone. For example, the site maintained by Pringles invites young visitors to "check us out."

Style and tone also depend on sentence length, figures of speech, and other linguistic choices, such as whether to convey your message in capital letters or to highlight key words and phrases. These choices put style and tone on a scale that ranges from formal to informal to casual. Once you know the specific audience you wish to target and understand the specific purpose or goal you wish to achieve, you will be prepared to adopt an appropriate style and tone for your web site. (For more information on style and tone related to effective web communication, see Chapter 1. Much of what applies to effective style and tone for email messages also applies to web communication.)

In sum, a web site that adopts a *serious* style and an authoritative tone will generally have these qualities:

- Be serious
- Maintain a figurative distance between the writer and reader
- Avoid personal references to maintain an impersonal tone
- Avoid contractions
- Incorporate abstract language in long sentences and paragraphs

A web site that espouses an *informal* style and a conversational tone—like the style and tone of this book—will have these characteristics:

- Show sentence variety
- Use first- and second-person pronouns and contractions to achieve a more personal tone
- Demonstrate a slight figurative distance between the writer and reader

Web sites that present a *casual* style and adopt a laid-back, cool tone can be described as follows:

- Use short- to medium-length sentences (20 words or fewer)
- Have short paragraphs (3 to 5 sentences)
- Incorporate descriptive language with specific words in conjunction with whimsical topics
- Use numerous personal pronouns and contractions
- Refer to people by first names, pet names, and nicknames
- Incorporate slang, jargon, and colloquialisms comfortably
- Minimize the distance between the writer and reader

Word Choice

Word choice also contributes to a web site's style and tone. Specifically, writers can choose active or passive constructions, everyday vocabulary or overblown language, succinct prose or excessive verbiage, and strong sentence beginnings or false subjects and starts. You'll find it helpful to explore each of these issues in a little more detail.

Active versus Passive Constructions

Active sentences are usually short and more direct than *passive* sentences. Active sentences—and active writing—are also more emphatic and confident. For these reasons, create active sentences when possible.

I suggest you create the initial draft of your web site's text without worrying about passive and active constructions. However, you'll need to learn how to spot passive constructions after you've written that first draft. Here's the formula:

Form of the verb *to be* plus the past participle of the verb = Passive construction

When searching for passive constructions, look for all forms of the verb *to be* and circle them. These forms include the following:

Am	Be	Is
Are	Been	Was
	Being	Were

(Don't confuse *have* or *had* with forms of the verb *to be*. They're not!)

For the next step, look at the verb that follows *to be*. If that verb is a past participle, you most likely have a passive construction. But what exactly is a *past participle?* It's the third column in any chart of verb conjugations.

For instance, if we conjugate the verb *complete,* we get these forms:

PRESENT	PAST	PAST PARTICIPLE
Complete	Completed	**Completed**

Even though the past and past participle forms of *complete* are the same, don't let that fool you. They won't always be the same. Just be sure to use the form in the third column, and you'll have the past participle.

Here's another way to determine the past participle of a verb:

I **complete** the test.

I **completed** the test.

I **have completed** the test.

The last sentence includes the past participle of *complete.* The past participle always goes in a sentence with either *have, has,* or *had.* But to determine whether the construction is passive, we need to leave the *have* behind. (Stay with me now! I know this discussion about passive constructions can be a little confusing.)

Now let's look at a sentence that takes a form of the verb *to be* and adds the past participle of *complete.* Here's the result:

The test **was completed** this morning.

The construction *was completed* is passive because it follows our passive formula:

To be + Past participle = Passive

If you see a sentence like this in an early draft of your web site's text, revise it. But how do you change a passive construction to active? First, you ask, Who completed the test this morning? Let's say the person who completed the test this morning is Miss Raven. Place *Miss Raven* in the subject position of the original sentence, and you'll have converted it from passive to active:

Miss Raven **completed** the test this morning.

Note the tense in the active sentence about Miss Raven is past tense. That's because the tense of the verb *to be* in the passive sentence is also past tense. Here's another example:

Traffic on I-95 **is slowed** every afternoon.

The form of the verb *to be* is present tense. And you can see that the past participle of *slow* follows it. Converting this passive sentence to an active one results in the following:

The Pennsylvania Highway Patrol **slows** traffic on I-95 every afternoon.

Notice that this revision to active voice is in present tense—*slows traffic*—because the form of the verb *to be* in the passive sentence is also present—*is*.

We're just about finished with this discussion about passive voice. I have a word of help and two warnings. Most verbs are called *regular* because in the past and past participle forms, they end in *-ed*. The past participle of *talk* is *talked*. The past participle of *study* is *studied*. To simplify our passive formula even more, look for *to be* followed by a verb that ends in *-ed*. Where's the passive in the following sentence?

Results were determined in the national laboratory last year.

Remember, find the form of the verb *to be* and the verb that ends in *-ed*. It's fairly simple, right?

Here are my warnings: First, don't allow any intervening words to trip you up. For example, if I were to include a modifier in the sentence above, I might get a sentence like this one:

Results were **quickly** determined in the national laboratory last year.

With or without the intervening *quickly,* the sentence is still passive.

My second warning is a bit more complicated. Remember when I said most past participles usually end in *-ed*? Unfortunately, *most* doesn't mean *all*. English includes its share of irregular verbs. With these verbs, the past participles—even the simple past forms—have endings other than *-ed*. These endings include *-t, -n, -k,* and *-d.* Here are some examples:

PRESENT	PAST	PAST PARTICIPLE
sell	sold	sold
buy	bought	bought
take	took	taken
drink	drank	drunk

Look at these four sentences, which use the past participle forms of these four words:

The soda was **sold.**

The soda was **bought.**

The soda was **taken.**

The soda was **drunk.**

By answering the question Who? for each of these four passive statements, we can easily convert them into active statements:

Paul **sold** the soda.

Paul **bought** the soda.

Paul **took** the soda.

Paul **drank** the soda.

Let's look at our passive formula one last time before we move on:

To be + *-ed, -d, -t,* or *-k* = Passive

Passive constructions aren't against the "grammar laws" that drive most English teachers crazy. In fact, you can and should use a passive construction when you don't know or don't want to reveal the true subject. Read this sentence:

Paper clips were dropped in the copy machine.

The person who dropped the paper clips is obviously avoided. If the receiver of the action is more important than the actor, passive is OK. Look at this sentence:

Your rebate check is attached.

If this information were included in a letter to me, I'd certainly be more interested in the *check* than in the *We*, as in *We attached your rebate check.*

So now that you have permission to use a passive construction, please remember: *Active sentences are preferred.* Oops. That's passive. Let me try again: *Readers prefer active sentences.*

Everyday Language versus Overblown Vocabulary

Most of us were taught in high school that having a sophisticated vocabulary was a sign to the world that we were educated. Just look at the standardized tests we take to get into colleges, graduate schools, law schools, and medical schools. The SAT, GRE, LSAT, and MCAT all have sections that test our knowledge of obscure vocabulary. While I'm sure there is some pedagogical justification for knowing bloated language, when you create web sites, simply unlearn what you learned about using an impressive vocabulary. With today's massive amounts of information and a common condition known as *information overload,* shorter is definitely better.

Are you still skeptical? Several organizations have adopted policies in light of the *plain English movement.* The Securities and Exchange Commission (SEC) now requires that annual reports be written in everyday English. As early as 1983, the U.S. Navy distributed lists of simple substitutes for overblown language, with the strong suggestion that people use the simpler versions. And a British-based group known as Clarity has documented numerous cases in which using plain English saved private and government organizations thousands of dollars.

Because I strongly believe that the purpose for any message—electronic or otherwise—is to get results, the quicker your readers understand your message, the quicker you'll get results. So avoid overblown language. Table 2.1 lists a few examples of overblown language and their everyday substitutes. While I agree that words such as *utilize* are perfectly good, why spend your space and your readers' time on three syllables when one is plenty? (I was just about to write "when one is sufficient." But I caught myself, which leads me to my final point about overblown vocabulary.) Any word that's made up of three or more syllables is a possible candidate for the chopping block. *Sufficient* has three syllables; *plenty* has only two. If you can think of a one-syllable synonym for *sufficient* or *plenty*, YOU should be writing this book!

TABLE 2.1   Overblown Language and Succinct Alternatives

Overblown Word	Better Alternative
accordingly	so
commence	begin
curtail	shorten
demonstrate	show
encounter	meet
fabricate	make
inform	tell
locate	find
manufacture	make
necessitate	compel
proceed	go
ramifications	results
sophisticated	complex
terminate	end
utilize	use
visualize	picture

Let me include one additional comment that's somewhat connected to overblown language: Overblown language makes text longer than necessary. Another technique for pruning your prose is to use one-word verbs instead of phrasal verbs, if possible. For example, write *bear* instead of *put up with* and *respect* instead of *look up to*. In addition to tightening your writing, this technique will help nonnative speakers of English who visit your site understand it.

Redundancies

In writing prose for your web site, avoid repetitiveness, verbosity, wordiness, windiness, prolixity, circumlocution, and verbal effusion. (I know, I know; I've violated the big-word rule.) Anyway, don't repeat yourself and don't qualify words or phrases unnecessarily. For instance, *large in size* is redundant because what else is *large* but a size? What about *past experience* and *actual experience*? They're redundant because you can't experience something in the future, and you can't experience something if it doesn't exist. How about this one? *Free gift.* When you go to the cosmetic counter in an upscale department store, you'll find various brands offering samples to promote their products. When they advertise those samples as *free gifts*, I go crazy. What is a *gift* if it isn't free?

While you should check a style guide or writing handbook for an exhaustive list of redundancies, I offer a few examples in Table 2.2 (page 78). Sensitize yourself to redundancies. Doing so will make the editing process one of common sense rather than rules.

False Subjects and False Starts

False subjects begin with *it* or *there* and typically end with *is* or *was*. Take a look at this sentence:

> It is because of Princess Bubba that visitors to St. John's Inn eat no bacon.

It is adds nothing to the sentence. Rather, beginning with *because* makes the sentence more concrete. Take a look at the sentence without the false subject:

> Because of Princess Bubba, visitors to St. John's Inn eat no bacon.

TABLE 2.2   **Redundancies and Suitable Alternatives**

Redundant Phrase	Alternative Word
absolutely complete	complete
basic fundamentals	fundamentals
check up on	check
disappear from sight	disappear
each and every	each
few in number	few
hopeful optimism	optimism
important essentials	essentials
joint cooperation	cooperation
mix together	mix
new innovation	innovation
one and the same	the same
period of time	period
repeat again	repeat
same identical	same
total of ten	ten
unsolved problem	problem

The revision here is shorter and more direct. (By the way, Princess Bubba is a Vietnamese pot-bellied pig who lives in Myrtle Beach, South Carolina. Her owners paint her "piggys" hot pink.)

Also be aware that false subjects can appear within sentences. Look at this sentence:

Annalee Otac decided **there was** some ham that she'd like to eat.

Here, the writer embedded the false subject within the sentence. Look at this revision:

Annalee Otac decided she'd like to eat some ham.

Again, the revision is more direct than the original version. Eliminating false subjects simply requires you to find a substantial noun after the false subject and let that word start your clause. For example:

There may be some products you wish to order from our web site.

We know *there may be* is a false subject because we can substitute *there are*. Because it's hard to revise the sentence to start with *some products,* you'll have to move to the next noun—*you.* Here's the revision:

You may wish to order some products from our web site.

The revision is more succinct, and it starts with a word more important than *there may be.*

Content

In conjunction with being sensitive to your word choice, your purpose, and your audience, you need to be sensitive about your web site's content. Because the web is a medium that can convey information through text, graphics (both moving and still), and audio, the decisions you make about content are complex. We'll spend most of the time in this section talking about textual considerations for your web site; we'll consider graphics and audio only briefly.

The act of reading text on a computer screen is very different from that of reading text on a printed page. First, the physical sensation is different. Electronic texts tie us to computer screens at desks that are less comfortable than reading chairs. In addition, we tend to be farther away from computer screens than from printed books and newspapers when we read. Thus, we read electronic information about 25% slower than printed material. As a result, we tend to become fatigued more quickly.

So, if you want your site to communicate your message effectively, you should write 40% to 50% less for a web site than you would for the printed page. Reducing text by this amount will help ensure your audience is physically comfortable while visiting your site—which probably means they'll stay there longer. People have short attention spans and want information as quickly as they can get it. Having a shorter text allows your readers to get through the information more quickly and more comfortably, as well.

The psychological sensation of reading an electronic text is also very different. With a printed text, you can physically see and feel the entire document. When you read a chapter in a book or an article in a magazine, for instance, you know what you're in for as you read chunks of text from page to page, beginning to end, usually in a linear fashion. You might justifiably argue that you jump around in magazines. But I would counterargue that when you find an article that truly holds your interest, you read it from start to finish—that is, if you want it to make sense.

Unlike that of a book or magazine, the wholeness of an electronic text isn't immediately clear. Moreover, studies have shown that people don't like to scroll through a lot of electronic text. If your visitors can't immediately grasp or visualize the length and organization of your web site—even a portion thereof—chances are they won't visit it for very long. Based on the physical and psychological sensations associated with reading electronic text, your site will be more effective if you limit the amount of textual information in it.

When you're faced with having to convey a great deal of information on your site, you'll have to decide whether to construct a series of related, short web pages or to present a lot of information on a few web pages. As a general guideline, long pages on web sites are acceptable if you think your audience will use the information the way they use it in a printed document. That is, if you expect your audience to download and print the information for use as a reference tool, long pages are more efficient because they use less paper and are easier to use in a printed format. There's still a caveat, though. Web sites with long, complex pages may take a long time to download and print, causing impatient readers to go elsewhere.

Your other option is to divide long, complex information into several short pages. Just be sure to focus each page on one subject, which creates an organization your audience can follow easily. The primary benefit of organizing information this way is that visitors to your site can download information quickly. The drawback is that printing numerous short web pages can be more tedious and use more paper.

As a general rule, don't be so enamored with the technology that you create more pages than you really need, especially when the information isn't complex. Put another way, don't divide information among several pages just because the technology allows you to do so. It's fine if you're the *Salt Lake Tribune* and have separate pages within your web site, such as "World," "Utah," "Sports," "Business," and "Crossword Puzzle." It's in the paper's interest to segment this information. First, it offers visitors an easy way to navigate the site. They can see what the paper offers and jump right to the section that interests them. And second, doing so allows readers to select, download, and read only those sections of the virtual newspaper that interest them. But if you want your visitors to "click here" to see a photograph of your favorite pooch, then "click here" to read your favorite poem, and then "click here" to hear your canary sing, you're simply gratifying your own love for the technology.

Some web sites give visitors choices for navigating. The "privacy" page maintained by the Social Security Administration allows users to view the site either by means of links, designed to make navigation easy, or through a single file version, designed to make printing more convenient.

Unfortunately, there's not a hard-and-fast rule for the ideal number of pages for a web site. Rather, you should allow how you expect your audience to use your site to determine how you organize its text among virtual pages. Reference manuals and narratives tend to include more *text* per page than do sites devoted to cybercommerce. Cybercommerce sites reserve space for photographs of their merchandise. Just remember: A site with a few long pages will probably take more time to download but will need less paper, and a site with a lot of short pages will probably be quicker to download but will need more paper. Whichever option you choose, be sure every page is there for a solid reason. If splitting up the information among multiple pages hinders, rather than enhances, readability, don't do it. After all, some web sites work just fine with one page.

Establishing and Maintaining Credibility

With the right equipment and technical know-how, anyone can create a web page and disseminate information to a wide audience at a relatively low cost. Think about the implications of that statement for a moment! Without an editorial process that makes writers responsible for everything they write, the web is saturated with all kinds of information from all kinds of sources—credible and not credible.

Certainly, when searching for information on the World Wide Web, you have to be a critical reader because the web has spawned its share of swindlers and hustlers. The prevalence of shady web sites has led the Federal Trade Commission (FTC) to conduct "surf days" several times a year, during which FTC staff, attorneys-general, and others across the United States and around the world spend hours scanning the web for scams.

This trend forces you, as a writer of valid information for the web, to work extra hard to establish and main credibility if your web site is going to communicate information effectively. This quality is what Aristotle referred to as *ethos*—or the character of the writer—in his discussions of effective rhetoric more than 3,000 years ago. If, for example, you're the American Red Cross, you've established your credibility just by what you do. So when the Red Cross solicits donations of people's time, money, or blood, they can comply, knowing that the organization is legitimate. If you're Mercedes-Benz, you've established your credibility by what you make. If you're the Reverend Billy Graham, you've established your credibility by who you are.

But what if you or your organization aren't as recognized as the Red Cross, Mercedes, or the Reverend Graham? *HelensMelons.com,* for example, might be a perfectly credible site for botanists interested in the morphology of the summer fruit. But like most sites, that site will need to rely on these rhetorical strategies to establish and maintain credibility:

- Correct spelling and grammar
- Links to other sites
- Up-to-date information
- Effective prose

Correct Spelling and Grammar

Your spelling and grammar have to be correct, and there's no negotiation on this point—at all! I know, I know. I sound like Ms. Fittich, the uptight English teacher. But truthfully, if you aren't going to take the time to check your site for spelling and grammatical details, what else might you be missing?

Checking your spelling is quite easy with the help of a spell-checker. However, eye think ewe should double Czech the spell-checker two catch what it Misses. A spell-check program will catch only words that are misspelled, not those that are used incorrectly.

Grammar checkers aren't 100% correct, either. That's because there are an infinite number of ways you and I can combine words into sentences. Because grammar-checkers can't anticipate all these combinations, they can't be programmed for all the possibilities. So you may create a sentence that confuses the grammar-checker.

Given this challenge, I suggest you ask someone else to review your text. Take a hard-copy draft of your web site text to your English teacher or writing professor at school. Ask for help from the communications staff in your office. Solicit input from a journalist you know. Or hire a communications consultant.

Links to Other Sites

The World Wide Web is a system that joins information through connections known as *links*. These links, for example, can connect web pages within sites, which are called *internal links,* or web pages among various sites, which are called *external links* (see Figure 2.2, page 84).

Highlighted or underlined words, phrases, or audio clips can serve as the icons that provide links on a web page. When you position the cursor over a link, it changes from an arrow to a pointing finger, which points you in the right direction. When you click the pointing finger on a link, the web takes you electronically to a different page within that particular site or to a different site altogether.

Think of it this way: When you open a book on, let's say, helicopters, you may jump to page 39 because you want to read about types of helicopters first. Once there, you may learn that page 43 has photographs of single rotor, coaxial rotor, and intermeshing rotor helicopters, and so you may jump there. On the World Wide Web, jumping from one link to another is the equivalent of jumping from one page or section to another in a printed book. If the helicopter book were to refer you to another publication and you were interested in getting that publication, you'd have to visit your local library or bookstore. However, the web gives you access to hundreds of thousands of sites, and you never have to leave your computer, your desk, or your chair.

FIGURE 2.2   Types of Links

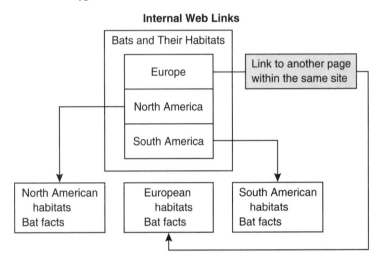

Internal Web Links

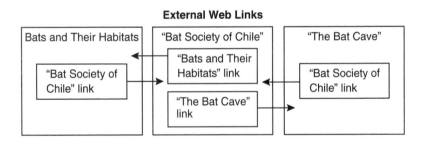

External Web Links

Note: Please don't assume this information contradicts the comment I made earlier about linear reading. It doesn't. The earlier point focused on a reader's limited ability to conceptualize a web site's organization and suggested that limiting the number of pages and amount of information contribute to a web site's efficacy.

So, how do links contribute to a site's credibility? Appropriate links create credibility for several reasons:

1. You demonstrate that you know your field.
2. You prove that you've done your research.
3. You're aware of other sites that complement the information on your site.

For example, The American Civil War Homepage, first launched in February 1995, started as a class project for an information science course at the University of Tennessee. It currently offers 21 pages of links to sites on topics as diverse as Civil War poetry and music, soldiers' photographs, nineteenth-century letters, and the 27th Pennsylvania Infantry Regiment. As a result, Civil War enthusiasts, scholars, and historians all find the site invaluable.

You'll probably avoid links to other sites, however, if your purpose is to convey information in a more proprietary manner and keep readers at your site. This means you'll have to rely on other methods to establish and maintain your credibility. The site maintained by the *New York Times*, which seeks to promote its publication through its information only, provides internal links to local, national, and world news. While you sense you're moving beyond the site as the links broaden, the links are exclusively within the *New York Times* web site.

If the purpose of your site is to sell a product or service, you'll also probably avoid links to other sites. Rather, you should consider including some or all of the following internal links: a company mission statement, company history, business philosophy, product line, services offered, technical support available, news in the industry, how to order products or services, and anything new. For instance, Phar-Mor, the discount food and drug store, includes internal links to "Health Tips," "Beauty Tips," and "Seasonal Tips," which the company uses to promote itself. Mt. Olive Pickle Company's site includes only internal links. One of Mt. Olive's internal links—"Great Moments in Pickle History"—tells visitors that Queen Elizabeth I (1533–1603) was a pickle fan. With this technique, Mt. Olive Pickle Company promotes itself through popular pickle trivia.

Used strategically, links can help you establish and maintain your credibility. But no matter what you establish as your purpose, be sure you've achieved your purpose with your audience before you send them away to other sites.

Up-to-Date Web Sites

Keeping your web site current is another technique that will help maintain its credibility. On a very basic level, a current web site demonstrates it's continually evolving in response to the changing education, legal, technical, medical, and business environments that interest your audience. A current site also demonstrates you're aware of how your site fits into the context of the present.

It's particularly important to maintain a current site if it purports to distribute up-to-date information. For example, you'll need to update your site with the seasons if you're selling the latest fashions from the runways of Paris and Milan. You'll need to update your site weekly if you're selling real estate, given constant changes in the market. And you'll need to update your site several times a day if it offers information on airline flight arrivals or regional weather patterns. The Alaska Region Headquarters, part of the National Weather Service, offers an hourly summary of weather data. Indeed, if your site is anything less than current, your visitors will go elsewhere for relevant information.

Keeping your web site current has an added advantage: It keeps people coming back again and again. If your site has the reputation of constantly changing, you give people reason to visit it again. The site maintained by the Edgar Allan Poe Museum in Richmond, Virginia, informs visitors, "We'll be updating this page from time to time . . . so come back and visit us often." If you have the resources to redesign your site every few months, you'll be able to take advantage of the latest technology.

At any rate, don't assume that you can create a web site once and be forever finished with it—unless you want your site to gain a "cob web" reputation. No doubt, effective web sites are always works in progress.

Skimmable Text

Links and updates aside, good writing is perhaps the most important strategy you can adopt for establishing and maintaining your web site's credibility. In their 1998 study *Applying Writing Guidelines to Web Pages,** John Morkes and Jakob Nielsen determined that 79% of web site

---

*Available at http://www.useit.com/papers/webwriting/rewriting.html.

visitors tend to scan or skim electronic text, rather than read it carefully. So if you want your visitors to perceive your site as credible, you'll need to offer them concise prose, or *skimmable text.* (See also the section titled Style and Tone, pages 68–81.)

One technique for achieving skimmable text on your site is to limit the number of words per sentence to 20 or fewer. If you're writing for an international audience whose primary language is not English, you should limit the number of words per sentence to 15 or fewer. While it's tedious, a straightforward method for diagnosing sentence length is simply to count the number of words per sentence.

Short paragraphs also contribute to skimmability. Write paragraphs in small chunks, and include only about half the words you'd include in a conventional printed document. Web site paragraphs, like printed paragraphs, should contain a main idea conveyed in one, two, three, or four sentences. Because short paragraphs allow your readers to focus easily on the main idea, they are quite effective.

To make your paragraphs even more effective, have your first sentence contain the *conclusion,* or the primary point you wish to make. This strategy is important because web pages load from top to bottom, and it's human nature to focus on the first bit of information we see. In your English class, you probably learned that topic sentences can go at the start, in the middle, or at the end of a paragraph. While that's perfectly acceptable in paragraphs written on paper, that's not effective for paragraphs written for web sites. Take a look at any commercially successful site, and you'll notice that the first graphic you see is a spiffy advertisement—paid for by the organization selling the product or service. We can learn from these professional web advertisers and mass marketers: Hook your visitors early, and you will have a greater chance of keeping them.

One last argument for putting your topic sentence at the top of each paragraph: Web visitors are impatient and don't like to scroll, so the information at the top of a site will get the most attention.

To keep your paragraphs unified, follow your conclusion (or main point) with important support material. Any necessary background information can go later. For a good example of this organizational technique, look at any newspaper story and analyze its organization. You'll see that the most important paragraph is the first one, with support provided by more and more specific details in succeeding paragraphs. Because newspaper publishers know that most readers won't turn to another page to finish reading a story (like web visitors who won't scroll or jump to another link in a site), they require reporters to offer readers the main point in the opening paragraph or two. Use that organizational technique on every page in your web site. Here's an outline for effective web site text organization:

- Conclusion
- Supporting point 1
- Supporting point 2
- Supporting point 3
- Background information

Finally, using effective emphasis techniques will strengthen the skimmability of your web site's text. Including headings, bullets, highlighted key words and phrases, and varied font sizes and types will attract the attention of your readers as they scan your site.

The use of headings can help readers skim your web site if you separate the headings slightly from the text to which they refer. Also make sure headings are succinct, specific, and not overused. Headings such as "Making Peanut Suet" and "Hanging Your Suet" on your "Suet Recipes" web site are likely to attract my attention, making it easier to find these topics than if they are buried in a paragraph. Using bulleted lists will also help me scan your site. For example, if you convey the recipe for peanut suet in traditional paragraph format, I'll find it harder to determine what ingredients I need than if you provide them in a bulleted list.

Highlighting key words and phrases also will attract your readers' attention, as long as the highlighted information provides enough of a context for your readers. Look at these two examples:

SENTENCE 1

Boil the **peanuts** for 45 minutes or until the mixture has absorbed all the water.

SENTENCE 2
Boil the peanuts for 45 minutes or until the mixture has
absorbed all the water.

The use of bold type in the second example helps you pick out the most
important information.
   Don't, however, overhighlight your information. Doing so will have
the opposite effect, as in this example:

Boil the peanuts for 45 minutes or until the **mixture has
absorbed all the water.**

Varying font type and size can also enhance the skimmability of your
site. But as with highlighting, don't overdo it—unless you want your web
site to look like a pasted-together ransom note! Here's the effect:

Downy Woodpeckers LOVE to dine on fresh suet
from the local butcher!

For emphasizing ideas, some web designers like to incorporate text
written in all uppercase letters (an email technique known as *flaming*).
Text written in all uppercase letters is not only difficult to read, but it
may also hide acronyms. Look at the following example:

JOAN JOINED NOW.

This sentence may prompt you to ask, When? But in fact, the question
should be What? Look at this revision:

Joan joined NOW.

In the first sentence, you might have mistaken *NOW* for a temporal ad-
verb, rather than the acronym for the National Organization for Women.

A final thought on techniques that enhance scannability: Consider your emphasis options carefully, and use them with moderation. Doing so will contribute to creating a wonderfully skimmable web site.

Web Site Construction and Navigation

To begin constructing your site, you won't need your computer. Rather, you'll need three old-fashioned items:

1. A stack of colored index cards
2. A large sheet of paper
3. A pencil or pen

First, assign a color to each level you expect your site to have. Your main page—also known as your *home page*—might be blue, your links yellow, and your sublinks pink.

While you can create more and more levels within your site, it's best to stop at three. Generally, no page on your site should be more than three clicks away from any other page. Anything more will cause cognitive overload for your visitors. And if they become psychologically lost in a site due to its complexity, they'll loose patience with it and leave.

It doesn't matter what colors you assign each level, just as long as the colors for each level are different:

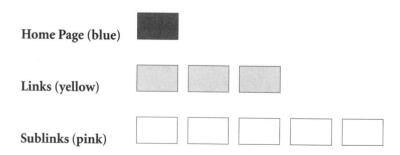

Home Page (blue)

Links (yellow)

Sublinks (pink)

Write a working title for your site on the blue card—the main card—and place it in the center of a large table. Make the title as specific as possible. Begin with a simple subject; then add a modifier or two. Let's go back to our "suet" example. I may create the first-draft title "Bird Suet" and then add the qualifying noun "Recipes" and the modifier "Easy." The result: "Easy Bird Suet Recipes." Because the main page, or home page, is the first page your audience sees when accessing your site, the title is extremely important. However, because the content of your home page depends on the content of the entire site, you may need to revise the title once you've created the entire site.

There's a very important reason to write the most strategic web title possible. When you publish your site, you'll have to register your home page, or *index page,* with search engines, such as AltaVista, Excite, Hot-Bot, Infoseek, Lycos, Magellan, Yahoo! and WebCrawler. These search engines depend on key words in titles to locate specific sites. Indeed, if you're going to all the trouble to make your site effective, you'll also want to make it findable. So make your title as specific as you can.

On a related note, the number of hits a search engine returns can literally run into the thousands. You may want to hire a consultant or schedule an appointment with your organization's technology specialist to learn about strategies related to your site's title that will display it as a top listing.

Let's now assume you've done everything possible to create a specific title that search engines return as a top listing. The next step, then, is to think of your home page as the front door to your house, school, or business. From that place, visitors can get a peek of what's inside and decide whether they want to go in. That's why your home page needs to include not only a descriptive title but also an overview of the site and navigational bar.

To create your site's overview and navigational bar, list on the sheet of paper all the topics you think would make effective links. Think of this step as a brainstorming session. Once you've come up with a list, transcribe those topics onto the yellow cards (one topic per card), and place them around the blue card on the table. Next, list on the sheet of paper the subtopics that relate to each of the topics on the yellow cards. Transcribe your subtopics to the pink cards (again, one per card), and place them around the appropriate yellow cards on the table. Having completed these steps, you're well on your way to developing an organized web site.

To create an effective overview for your site, look back at the sheet of paper on which you listed topics and subtopics. Based on that collective information, create a rough paragraph that provides an overview of the site. Add appropriate emphasis techniques and transcribe that information onto the blue card. Once at your site, your audience will appreciate the succinct overview you provide to help determine the site's relevance to their interests. The overview on the site maintained by Abercrombie & Fitch, for example, welcomes "fun seekers [who] need look no further than abercrombie.com" for the latest young adult fashions.

In addition to a specific title and overview on your site's home page, you'll need to include a virtual table of contents, or *navigational bar*. The navigational bar gives your readers a virtual map of your site, including links, so they not only know the content of your site but also how to retrieve it. The site operated by the Utah Jazz, a professional basketball team, offers a navigational bar that includes links to scores, players, statistics, and tickets. On some web sites, you'll see a graphical navigational bar in addition to the textual navigational bar. Users can select links in either way, depending on their software and preference.

To create a working navigational bar—the icon on each page that serves as a table of contents and guides visitors around your site—transcribe the topics from the yellow cards to a single list on the blue card (see Figure 2.3). Then transcribe the working navigational bar to every

---

FIGURE 2.3   Home Page Card (Blue)

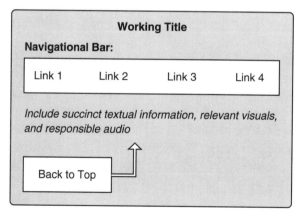

FIGURE 2.4   Topic Card (Yellow)

FIGURE 2.4   Topic Card (Yellow)

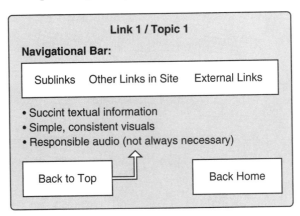

yellow card (see Figure 2.4). Include *"Back Home"* on every navigational bar except the one on your home page. Most navigational bars appear at the top of the web site. So it's a good idea to include *"Back to Top of Page"* on every page. If some of your pages run long, you'll want to make it easy for your visitors to get to the top of each page and its navigational bar. If you decide to create navigational bars linking your topics to your subtopics, just follow the same technique as you did for your primary navigational bar. Figure 2.5 (page 94) shows how the cards will look once you've completed this organizational exercise.

Web Addresses

When you want customers to find your store or friends to find your house, you probably give them directions using your address. To get customers and friends to your web site, you'll do the same.

First you'll need to create an URL, or *Uniform Resource Locator*. It's the series of characters, also known as a web address, that identifies each specific page on the World Wide Web. Take a look at the following URL for the White House in Washington, DC:

http://www.whitehouse.gov/

FIGURE 2.5   Organizing Your Web Site

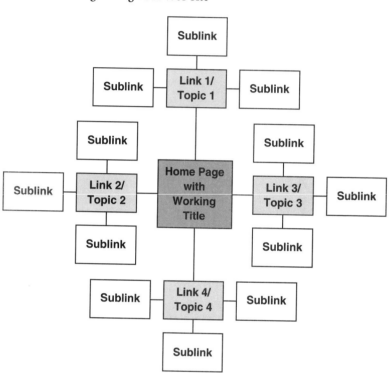

The *http://* stands for *Hypertext Transport Protocol.* It's a communication system that browsers and servers understand, and it moves web pages from computer to computer around the World Wide Web. The *www* in the address simply refers to the World Wide Web, while *whitehouse* refers to the name of the server that hosts this site. The domain name is *gov* and indicates that the server is associated with the U.S. government.*

Some organizations have their own servers, which they name to reflect their companies, products, or services. The server that handles the site maintained by the makers of hand-baked Moravian Spice Cookies is named *salembaking* to reflect the North Carolina city where the cookies

*If you're interested in learning more about web addresses, spend some time with the glossary on the InterNIC site. You'll find that web address on page 95 and in the list at the end of this chapter.

are made. The *com* portion of the address, the domain, indicates the organization operating the server is a commercial one:

http://www.salembaking.com/

But be warned: Special server names can come at a high price. The server name *computer.com* is apparently so coveted that its street value is currently $500,000!

If you don't have your own server, your address will have to reflect the server that hosts your site as well as the type of organization that operates the server. When you have a choice, select a concise, sensible web address. This is important because the only way people will be able to find you (if they haven't been referred to your site through a search engine or external link) is to type your address into their web browser manually. So make it easy on your potential visitors.

If you have the option to choose a name for your server, consider one that advertises not only your site but also your company, product, service, or viewpoint. For example, you might name your server after any of the following:

A company	http://www.sears.com/
A brand	http://www.wrangler.com/
A service	http://www.redcross.org/
A specific viewpoint	http://ww.pro-life.org/

Thinking back to the specific purpose of your site can help you choose an effective name for your server.

Keep in mind, however, that not every address you consider will be available. You'll have to try your proposed address at the InterNIC interface. Operated in cooperation with the National Science Foundation and Network Solutions, Inc., InterNIC is a registration services database. The site offers a useful tutorial to guide you through selecting a web address for your site. You can find InterNIC at this address:

http://www.networksolutions.com/cgi-bin/whois/whois

Let me say a few more words about the final element in a web address, the *domain*. It identifies the type of organization operating the server. As with email, some common suffixes include (but aren't limited to) the following:

Domain	Example
com (**com**mercial)	M&Ms http://www.m-ms.com/
edu (**edu**cational)	University of Hawaii http://www.hawaii.edu/
gov (**gov**ernment)	Social Security Administration http://www.ssa.gov/
mil (**mil**itary)	U.S. Navy http://www.navy.mil
net (**net**work)	American TeleSource International http://www.ati.net
org (nonprofit **org**anization)	The Nature Conservancy http://www.tnc.org/

Once you've created, established, and registered your web address, make sure every visitor to your site knows your address. For that reason, include the address on every page of your web site (even though certain browsers automatically print the web address on each page). Include an explicit statement, such as *The web address for this site is http://www. movies.com/.* If you include an explicit reference to your web address, you'll assure those who want the address can get it easily.

One last note about publicizing your web address: Not everyone will know to look for your site on the web, so be sure to print your web address on all your literature, stationery, advertisements, and business cards.

Design and Layout

Effective web design and layout require that you display your information simply and tastefully. Too much visual complexity can cause distractions that effectively bury your message. As a general rule, remember that "Less is more" for web sites.

Strive for consistency among web pages by creating a template or using a style sheet for each page. Templates and style sheets establish visual continuity, the virtual equivalent of brand awareness, for visitors to your site.

A consistent theme or graphic that unifies the entire site, such as a catchy phrase or company logo, can also provide a sense of familiarity for visitors. The official site for the United States Postal Service includes its logo on each page:

While graphics such as this are effective for unifying web sites, using too many can make a site look cluttered and dilute the focus of your message. If the graphics compete with one another for attention, your message may get lost in the mayhem.

To be effective, graphics should direct your readers' eyes to the most important information, something that's more easily accomplished with fewer graphics. Think for a moment about people who wear rings. When are you more apt to notice their glittery, bejeweled fingers? Probably when they're wearing one or two rings instead of eight or ten. Graphics are the same way.

Chose color combinations that enhance readability. The sharpest contrast possible is the most desirable. At the most basic level, select a dark background for light text and a light background for dark text. If you're unsure about the color combination you've chosen, try the *squint test*. Squint your eyes at the screen; if the text or the images fade into the background, then the contrast isn't sharp enough.

Once you've decided on a combination of colors, be consistent among the pages to unify your site. While changing colors may entertain *you*, it may distract or annoy your visitors, perhaps taking them back to the psychedelic days of the 1960s. Also remember that not all monitors have 256-color capability. Because many still have only 16 colors, test your site—graphics and all—on a monitor with such limited color capabilities.

Wallpaper choices are related to color choices. *Wallpaper*—which is the background design of your site—comes with web design software, or you can design it yourself. Just be sure not to use complex patterns that will distract your visitors or make reading difficult. Since the newness first associated with wallpaper variations has subsided, most web designers choose simple white backgrounds with dark text. This selection has the added bonus of being readable by people using the widest range of computer equipment.

Font types and sizes should complement, not complete with, your site's background. Limit the number of variations. Again, consistency enhances readability. Why challenge your audience to spend valuable time sorting out layers of different type? Instead, help them concentrate on your message. While you don't need to restrict yourself to just one font, keep font choices consistent among pages and use variations to emphasize specific points. For most sites, two fonts will suffice—one for headings and one for text.

Whenever possible, choose *sans serif* fonts, those without the little feet. They convey a crisp appearance on screen:

Geneva   Avant Garde

Helvetica   Gill Sans

Serif fonts, such as Times New Roman and that used in this book, have little feet on the ends of the letters. They're especially effective in printed documents because they help lead readers from one word to the next. However, when conveying information via a video screen, as you do with a web site, the distance among the pixels on the screen can have a fuzzying effect on serif fonts.

With web sites, you also have the option of including automatic scrolling, fading, blinking, and spinning features. However, because people's eyes are attracted to motion, using too many of these options can take attention away from your message. Use these options sparingly. If you choose to include motion on your site, make the text or graphic as small as possible, place it in a corner, and limit the number used to one. Clearly, from a design and layout standpoint, don't overdo anything when creating a web site.

Web sites today contain *frames*—separate windows within a web site's main window. For example, a site may include the title and overview on the main page with a navigation bar off to the left in a separate frame formed by means of a border or box. Frames that enhance readers' navigation around the site are useful. However, as with graphics and background images, don't overdo it. Keep in mind that frames can complicate a user's ability to conceptualize and navigate a site. Also, certain browsers can't display sites with frames. One study put the figure at 13%, which means you'll lose that percentage of visitors to your site if you incorporate frames.

Frames can also cause problems with the "Back" button, disabling it in some browsers. Because the "Back" button takes users back to the starting point, when it doesn't work, users are forced to exit your site, making it virtually useless. If that's not enough of a deterrent, keep in mind that certain browsers can't print framed pages appropriately or at all, especially when the frames are scrolling frames. Frames can also create challenges for search engines because they don't know what part of the frame to include in the search engine index.

Tom Milazzo, who has directed several information and technical resources departments for major organizations, sees good and bad in frames. With frames, visitors who access your home page download everything on your site, which is efficient. The disadvantage is that users might directly access a page deep within your site—not your home page. When they do so, users have accessed the site without the frame—that is, without that additional page within the site. Because many web designers include navigation buttons on frames, users who access a page deep in your web site directly, without going through your home page, lose the ability to access whatever is on the frame. This means they have no way to navigate within the site. To illustrate the problem, Milazzo suggests trying *http://www.bschool.unc.edu*, the University of North Carolina business school's home page, and then exiting and trying *http://www.bschool.unc.edu/infocenters/news6.html*. Doing so will demonstrate navigation limitations.

A few more words about design: If possible, avoid counters, which indicate the number of visitors, or hits, to your site. If, for marketing purposes, you or your organization really need a counter, then hide it somewhere on the page. A large number of hits can make you appear overconfident, which is not effective. And a small number of hits can make your site look amateurish, which isn't good, either.

Also don't include "Under construction" signs. They indicate your site isn't really open for business, which may send your visitors elsewhere. In fact, if you're responsive to your audience's needs and wants, you'll continually be updating your site. As such, any effective web site undergoing revisions and additions is always under construction.

Once you've created a working draft of your web site, test it to ensure its effectiveness. First, you'll have to upload it, and second, you should try it on various computers. Because you have quite a few options for uploading your site, you may want to enlist the help of a web specialist, unless you're able to do it yourself. One option is to save your pages on a diskette and forward it to your ISP, who will upload the pages for you.

After you've uploaded the pages, take a look at them on various computers because memory, various color resolutions, and screen size variations will affect how a page looks. Test the pages with a text-only browser, such as Lynx, or with the images turned off. If you assume people will visit your site under ideal conditions or conditions that mirror your own, you'll be making a serious mistake. The point is to make your site appealing to as many people, with as many computers, as possible.

Privacy, Legal, and Ethical Issues

There is growing concern about privacy on the World Wide Web. This means you'll have to be clear about how you use information you gather from visitors, how you protect that information, how you protect your organization's privacy, and what laws affect privacy on the web.

How You Use Information

In early 1998, the FTC completed a study of web practices and determined that most web sites collect personal information and many sell that information to direct marketers. This trend has understandably caused concern among web users.

One common method by which commercial and public web sites gather information is through *cookies.* Cookies deliver small data structures or files to users' computers, which stay on them for future reference by the commercial organizations that put them there. When you revisit the web site of such an organization, it identifies you as a return visitor. Why is this useful? Put to use appropriately, cookies can provide valuable information about users' browsing and buying habits, serving as an invaluable tool for marketing strategies.

Cookies allow web sites to maintain information about visitors in much the same way frequent-flyer profiles work. Let's say you prefer having a window seat in first class. An airline that has software to process such specific information can retain that preference. Then each time you book a flight, the airline can accommodate your preference without asking you for the same information again. Similarly, cookies recognize visitors as repeat customers and serve up personalized information based on previous visits—perhaps even a personal welcome and advertisements that seem to be tailored just for an individual customer.

If you decide to include cookies on your web site, keep in mind that the information collected can be inaccurate, taken out of the intended context, or misused. Cookies identify visitors to a particular web site by name, by email address, by the length of time they stayed at the site, and by the subjects and products that interest them. So there's absolutely no doubt that cookies can threaten visitors' privacy. Some opponents claim that cookies foster the notion that "Big Brother" is watching and recording your every move. But if your purpose for using cookies is simply to refine your web site so it's responsive to visitors' needs, then you probably won't run into any ethical issues.

If, on the other hand, you want to track information about the people who visit your site, decide specifically what information you'll collect from them and what you plan to do with it. Also be clear with your visitors about what you will do with that information.

Today, most web sites include a section on privacy, spelling out these very issues. Before you create your section on privacy, specifically identify how you plan to use personal information. Will you use the information in any of these ways?

1. To fulfill a request for information?
2. To count the number of visits to your site?
3. To respond to a concern or complaint?
4. To measure demographic information about visitors?
5. To provide information to other companies or marketers?

Once you've clearly established how you plan to use information about visitors to your site, take the time to examine several privacy statements currently on the web. The privacy link on the Bank of America home page, for example, explicitly states, "We are committed to protecting the information you provide to us," and it assures visitors that the bank uses information gathered from web visits to refine its services. As an ethical web site owner, you are obligated to inform visitors of how you will use their information.

How You Protect Visitor Information

Letting your visitors know how you'll use their information is not enough. You also need to tell visitors how you plan to protect the data they transmit through your site. When the transmission involves sensitive data—such as personal information, credit card numbers, social security numbers, telephone numbers, and addresses—encryption software can help.

Encryption software scrambles information so only the intended recipient can read it. There are two types:

1. *Private key encryption* is a method whereby both the sender and receiver of information through a web site share a single, common code to scramble a message for transmission and unscramble the message once it's arrived at its destination.

2. *Public key encryption* is a method through which everyone can scramble information transmitted through the web. However, only a private key holder can unscramble a message. In this method, the public and private keys work together, which helps protect a message's content.

How You Protect Your Organization's Privacy

Protecting your visitors' privacy is only one part of the privacy issue. You'll also have to be aware of how you protect the privacy of your organization. Because anyone with your web address can access your site, you shouldn't post any proprietary or secret information on it. It's good practice to inform members of your organization who are familiar with corporate secrets to review any postings. Doing so will serve as a good filter before anything inappropriate is put on the web. Once you've posted information on a site, you can't get it back. You can only close the site, and sometimes the damage has already been done.

A company's new logo, strategic vision, restricted software, and customer database are all targets for web theft. One major car manufacturer was set to unveil a new model at an international car exposition. When pictures and specifications of the model showed up on the company's web site, executives treated the snafu as a major marketing disaster. One section of the web site maintained by the Social Security Administration mistakenly included taxpayers' earnings and benefits. After just two days of getting up to 80 hits a second, the administration had to close that part of its site because of the implications for ex-spouses and divorce attorneys. If you or anyone else can claim that the information on your web site is confidential, don't post it. It isn't secure.

Privacy and the Law

As noted earlier, you need to be sensitive to visitors' expectations and needs as well as your organization's privacy. You'll also need to stay informed regarding the law. Because the virtual landscape of the World Wide Web is changing so rapidly, it's difficult to discuss current laws related to web privacy. But if you are going to create and maintain a web site, you should do your best to stay up to date with online privacy issues.

The Electronic Privacy Information Center is a useful resource that can help you stay current with legal issues. And the FTC conducts periodic public workshops that address electronic privacy issues, focusing on educating consumers and businesses about the use of personal information gathered online and encouraging consumer confidence through responsible online marketing practices.

Even so, lawmakers, the Department of Commerce, watchdog groups, and even online marketers continue to be extremely concerned about online privacy. In 1997, U.S. Senators Dianne Feinstein and Charles Grassley introduced the Personal Information Privacy Act after Feinstein's staffers were able to locate her social security number on the net in fewer than three minutes.

In response to growing government involvement in Internet and web site regulation, 51 privately owned businesses joined forces to create the Online Privacy Alliance in mid 1998. Their goal is to protect the industry's right to regulate its own privacy policies.

Memory, Load Time, Graphics, and Audio

Even though the primary focus of this book is on electronic *textual* communication, I'd like to spend a little time addressing memory, load time, graphics, and audio only as they relate to communication.

Sophisticated technology will probably tempt your creative powers as you design the graphic and audio components of your web site, but some of those choices will affect how effectively your web site communicates its message. While some web designers believe that professionally incorporated graphics and sound can enhance the credibility of a web site, be careful to avoid these two assumptions:

1. That just because you have state-of-the-art equipment, with a superfast modem and CPU and multimedia speakers, everyone else does, too.

2. That if a web page looks flashy on your machine, it will naturally look flashy on everybody else's machines.

As a responsible web page designer—whether you're just beginning or are more experienced—you can't make either assumption. Rather, you must ask yourself, How does my technology compare to everybody else's or at least to my target audience's technology?

Knowing something about computer memory and load time will help you make the best multimedia choices possible. As a general rule, professional web designers suggest that graphics should be no bigger than 30K to 40K; otherwise, your pages may take too long to load.

Memory

Many factors affect the speed with which web sites load:

1. The kind of operating system your visitors use

2. The video boards they have

3. The number and types of applications running at the same time

4. Monitor resolution, color choices, and color palette depth

5. RAM and video memory

For example, a computer with 16 meg of RAM (not a particularly underpowered machine) can experience problems with complex web sites, causing the computer to crash or hang. For that kind of machine, a web site with too many fancy graphics may only partially load. If you've written a web site that crashes or hangs the computers of one-third of the people who visit, your site won't be popular for long. For example, the site for the Chicago White Sox—with its interactive scoreboard and interactive roster—is great. But if your machine can't handle the site, the experience will be disappointing.

Load Time

No doubt, we live in an impatient, results-oriented world, and we transfer that impatience to our computers by demanding that web sites download quickly. In fact, web users need some information in less than one second if they are to move from page to page efficiently and comfortably. Thus, in a very real way, speed and efficiency contribute to a site's communication effectiveness. All your attention to systematic organization, snazzy graphics, and amplified audio won't mean much if you subject your audience to a long "World Wide Wait." They'll just go "webbing" elsewhere.

Test your site's load time with a 14.4 modem, the slowest model currently in use. Doing so will let you know how long people with the slowest Internet connections will have to wait to see your site. As a general rule, base your initial design decisions on the fastest load time possible on the greatest number and kinds of machines.

Having fewer graphics and audio files should speed up your load time but may bore your viewers. You'll have to determine whether visitors consider a simplistic web site too dull or a multimedia extravaganza too time consuming. The answer depends on your site's purpose and your audience's equipment.

When the purpose of your web site is to persuade people to buy something, you'll probably incorporate more graphics than when your purpose is simply to convey information. For example, if you expect visitors to shop on your site, you'll have to show them the polka-dotted men's shirt they're thinking of buying. The Gap—which sells an image as well as clothing to Generation X-ers—offers a high-quality web site that includes scanned-in photographs to promote the company's merchandise. In addition, the site offers an interactive page called "Get Dressed," which allows you to dress the site's virtual model with clothing you select. For that company and that site, graphics aren't only important but also necessary.

On the other hand, flashy logos, waving banners, navigational maps, and various sounds often don't contribute much to a site when its purpose is simply to convey news or academic information. After all, the ultimate product on such a site is information. The site for the *Chicago Tribune,* for example, allows visitors to choose a low-text or low-graphics version that provides links to "News," "Sports," "Business," and "Marketplace," with sublinks to "Nation/World," "Opinions," and "Weather."

Certainly, this simpler option speeds up the time you have to wait to download this site.

Again, give some thought to the kind of equipment your visitors use to guide the multimedia decisions you make. If you're creating a web site exclusively for an organization's employees, design the site to match the computing capabilities of the company's equipment. The Chrysler Corporation's intranet, for example, allows car dealers across the United States to bid on and buy used cars from "cyber car lots." Because that web site is limited to Chrysler employees, it's easy for computer technicians to design pages for the company's equipment. Similarly, the MacIntyre Business School at the University of Virginia provides faculty, staff, and students with access to the latest, fastest computers with a direct connection to the Internet. The school's web designers can thus create fancy multimedia sites for its intranet because they know users will be able to access the sites efficiently.

When the audience is external to the organization, web designers can use demographic data to speculate on the kinds of equipment visitors will use to access a given site. Indeed, if Rolls Royce decided to target its web site to affluent visitors, web designers at that organization could make the assumption that those visitors would have the latest computing equipment. Similarly, if Cadillac decided to target its site to older visitors, their web designers could assume that users are surfing the web with their son's or daughter's less-powerful castoff computers. General Motors, on the other hand, publicizes information that interests all kinds of people with a range of computing capabilities—from the newest, fastest computers to dinosaurs that are five years old and have slow modems. To reach the widest audience possible, General Motors would have to be conscious of this range when making multimedia decisions for its site.

Some big corporations and organizations, with large and varied cyberaudiences, continue to be responsive to accommodating people with older, slower computers by offering users a text-only option. That is, web sites give users the option to view the information without the fancy graphics, audio, or video. The site maintained by the U.S. National Park Service, for example, offers visitors a text-only option. Even so, as the population gets more powerful computers and web authors do more with their sites, text-only versions are becoming less common.

Perhaps the best way to accommodate most of your audience is to offer them the option for the format that's most compatible with their equipment, whether that's a high-graphics, low-graphics, or text-only version of your web site. People with graphic browsers—such as Netscape, Microsoft Internet Explorer, and Mosaic—can access the fancy graphics version, while those people with text-only browsers—such as Lynx and Gopher—can still access the textual information.

The IRS, for example, offers users a text-only version in addition to the graphics version, each a mouse click away. Travelocity, a popular site devoted to travel, offers visitors an easy version and a fast version. Visitors can choose high-graphics or text-only versions of the site.

If you decide to offer a text-only version of your site, consider labeling your images with the *alt* tag, an attribute used to supply textual information about a graphic. In other words, the alt tag describes the graphic. Using it is effective for visitors who use nongraphical or text-only browsers. And it's also useful for those who use Netscape, for example, with the graphics option turned off to speed download times.

If you decide to include your company's logo on each page of your web site and your visitors surf without a graphical browser, they'll only see a blank square with a small x in the center of a small box where the logo should be. Without the alt attribution, your audience will have no way of knowing what image goes there. With the alt attribution, however, your audience will see a short written description of the image— something like "Pfizer logo." That way, your audience will know what that space is reserved for. There is an added advantage to using the alt attribution: If your audience uses a graphical browser, they'll get a written description of the image as the page loads. Including the alt tag ensures that everyone who visits your site—whether with graphic or text-only browsers—will at least know what images are present and how they contribute to the overall purpose of your site.

Audio

Audio also requires careful consideration because today's computers are equipped to transmit not only music from the latest CDs but sound from web sites. You can't control the volume setting on your visitors' speakers, but you can choose the software that controls the volume of your sound files. You don't want to startle your visitors with a particularly jarring sound as they download your site. A site devoted to Edgar Allan Poe aficionados opens up with a crashing lightning strike and thunderclap that might scare unsuspecting visitors.

Volume isn't the only consideration regarding sound. Also consider the content of your sound files, which can unnecessarily scare visitors. One site includes a computer-generated voice that says, "Warning, virus detected; reformat your hard drive before it spreads." While the site's owner may have found the warning funny, it's really unethical to transmit lies in this way. Moreover, scaring people with lies and annoying them with nonsense is not an effective use of web space.

Some sites, however, use sound effectively. The web site maintained by Coca-Cola will make you want to stop what you're doing and get a Coke, even if you're not thirsty. It includes the sound of a carbonated beverage being poured into a glass, fizz and all. Sure, we've all heard that sound before, but it's nevertheless clever.

If you decide to include sound on your web site, you should test it on your friends, co-workers, or relatives. Their reactions will help you determine how your audience will react, and you can make decisions accordingly.

Writing is challenging in any communication medium, electronic or not. It's both a skill and an art. When you add web page design to the process, writing brings with it many choices and decisions. No doubt, a lot of brain work goes into creating an effective web site. But break the tasks into small units, take your time, get feedback, and practice. The results can be truly amazing!

Summary

Planning

- Establish the purpose of your website early.
- Analyze your potential audience.

Textual Elements

- Keep sentences, paragraphs, and pages short.
- Present your conclusion in the first sentence of each paragraph followed by support materials.
- Avoid redundancies and overblown language.
- Include everyday vocabulary.
- Use frames with caution.

- Create a consistent layout, including a navigational bar, "Back Home" link, and "Back to Top" link on each page.
- Sketch out your organizational plan before writing the HTML code.
- Incorporate no more than three levels of text within your site.

Emphasis Techniques
- Add adequate space between blocks of text.
- Include bulleted lists, highlighted words, and effective headings.
- Use both upper- and lowercase letters.
- Include a descriptive title.

Audio, Video, and Graphics
- Use the same graphics for each page to create consistency.
- Limit color combinations and font types and sizes, and incorporate them consistently throughout the site.
- Limit the use of motion.

Credibility
- Create links to other sites, keep your site current, and incorporate effective prose.
- Be sure spelling and grammar are correct.
- Incorporate an honest style, not a promotional one.
- Convey textual information in a positive tone; include what you *can* do, rather than what you *can't.*
- Avoid counters and "Under construction" signs.

Technology
- Strive for the greatest compatibility among the largest number of machines.
- Pay attention to memory and load time.
- Choose a web address that reflects your organization, that's professional, and that's easy to remember.

Legal and Ethical Issues
- Familiarize yourself with privacy laws and ethical issues.

Web Sites Consulted

Abercrombie & Fitch	http://www.abercrombie.com/
Alaska Region Headquarters	http://www.alaska.net/~nwsar
The American Civil War Homepage	http://sunsite.utk.edu/civil-war/aboutcwarhp.html/
American Planning Association	http://www.planning.org/
American Red Cross	http://www.redcross.org/
American TeleSource International	http://www.ati.net/
Amtrak	http://www.amtrak.com/
Anti-Gun Coalition of America	http://www.bitsnet.com/agca/
Bank of America	http://bankofamerica.com/
Beanie Babies	http://www.beaniebabies.com/
BellSouth	http://www.bellsouth.com/
Billy Graham	http://www.billygraham.org/
Bomb Pop	http://www.bombpop.com/
Bug Club	http://www.ex.ac.uk/bugclub/
Cable News Network (CNN)	http://www.cnn.com/
Cadillac	http://www.cadillac.com/
Cheap Tickets	http://www.cheaptickets.com/
Chicago Tribune	http://www.chicagotribune.com/
Chicago White Sox	http://www.chisox.com/
Coca-Cola	http://www.cocacola.com/
Dell Computer	http://www.dell.com/
Edgar Allan Poe Museum	http://www.poemuseum.org/
Electronic Privacy Information Center	http://www.epic.org/
Federal Bureau of Investigation (FBI)	http://www.fbi.gov/

Gap	http://www.gap.com/
Georgia Tech Web User Surveys	http://www.cc.gatech.edu/gvu/ user_surveys
General Motors	http://www.gm.com/
Internal Revenue Service (IRS)	http://www.irs.ustreas.gov/
InterNIC	http://www.networksolutions.com/ cgi-bin/whois/whois
Kenan-Flagler Business School	http://itr.bschool.unc.edu/ http://www.bschool.unc.edu/ infocenters/news6.html
M&Ms	http://www.m-ms.com/
Mercedes-Benz	http://www.mercedes-benz.com/
Movies.com	http://www.movies.go.com/
Moravian Spice Cookies	http://www.salembaking.com/
Mt. Olive Pickle Company	http://mtolivepickles.com/
Nabisco	http://www.nabisco.com/
National Public Radio (NPR)	http://www.npr.org/
National Rifle Association (NRA)	http://www.nra.org/
Nature Conservancy	http://www.tnc.org/
New York Times	http://www.newyorktimes.org/
Nielsen and Morkes	http://www.useit.com/alertbox/ 9710a.html/
Peanuts Comic Strip	http://www.snoopy.com/
Pharmor	http://www.pharmor.com/
Pringles	http://www.pringles.com/
Pro-Life	http://www.pro-life.org/
Public Broadcasting Service (PBS)	http://www.pbs.org/
Rolls Royce	http://www.rolls-royceandbentley. co.uk/

Salt Lake Tribune	http://www.sltrib.com/
Saturn	http://www.saturn.com
Schwab, Charles	http://www.schwab.com/
Sears	http://www.sears.com/
Social Security Administration	http://www.ssa.gov/
Tide ClothesLine	http://www.tide.com/
Travelocity	http://www.travelocity.com/
University of Cambridge	http://www.cam.ac.uk/
University of Hawaii	http://www.hawaii.edu/
University of Tennessee at Knoxville	http://www.utk.edu
U.S. Department of Education	http://www.ed.gov/
U.S. Department of Housing and Urban Development	http://www.hud.gov/
U.S. Navy	http://www.navy.mil/
Utah Jazz	http://www.nba.com/jazz/
WalMart	http://www.wal-mart.com/
Washington Post	http://www.washingtonpost.com/
White House	http://www.whitehouse.gov/
Wrangler	http://www.wrangler.com/

3

The Usenet

The Usenet is perhaps the wildest frontier on the Internet—a place where anybody who owns a computer and has the appropriate software can write messages and send them out into a public electronic forum or simply read those messages posted by others. Created in 1979 by two students at Duke University, the Usenet was one of the first tools available in the Internet, predating the World Wide Web by more than a decade.

The Usenet—an information-useful electronic network—contains more than 20,000 individual *newsgroups,* each focusing on a specific subject. While all computer users are welcome to browse through the messages written by others, it would be impossible to cover even a fraction of these messages. In fact, users post an estimated 100,000 messages to newsgroups each day!

To help you understand how a newsgroup works, think about its nonelectronic counterpart: an old-fashioned bulletin board that might hang, say, in an employee lounge, local community center, or college student union. On this hypothetical bulletin board, people can scribble notes on pieces of paper and pin them up. These notes can pertain to almost any subject. Some people may choose to write notes expressing their opinions on politics or the news of the day. Others could scrawl questions about problems they're facing at work. Garden enthusiasts might post notes sharing tips on how to grow more beautiful plants. Some people could even start to write long, fictional essays and post them on the bulletin board, only to have others add their ideas. And still other people might suggest that the company cafeteria offer organic popcorn or sun-dried tomato dressing.

No doubt, such a popular bulletin board would become cluttered and disorganized rather quickly. Given the vast amount and variety of material, it would make sense to organize the information into categories and make *different* bulletin boards. For example, people who wanted to post notes expressing opinions on the day's news would be encouraged to do so on a "Current Events" board. People with gardening information to share would place their notes on a "Gardening Tips" board. Still others would place their notes on yet other bulletin boards.

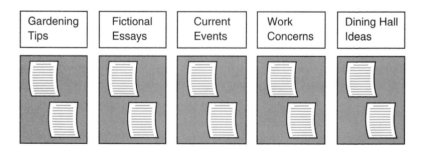

The Usenet functions like those segmented yet collective bulletin boards. When you look at the Usenet in this way, it's easy to understand why each newsgroup is called an *online bulletin board*. It's also easy to understand that the language of *posting a message to a newsgroup* comes from pinning up messages on bulletin boards. *Usenet* simply refers to the entire electronic system, *online bulletin board* simply refers to the individual group, and *posting* refers to putting your message out there for all to read.

Because the online community is so large, it's possible to find one or more newsgroups for almost any interest. Some newsgroups tend to appeal to people in certain professions. For instance, the group called *rec.radio.broadcasting* attracts mainly individuals who work in the broadcasting business. Accountants share tips of their trade in *alt. accounting*. Other newsgroups appeal to common interests. Baseball fans can discuss the fortunes of their favorite teams in *rec.sport.baseball*. People who like to exchange recipes can do so in *rec.food.cooking*. Home buyers can share information in groups such as *misc.consumers.house*, and viewers of the animated television show *The Simpsons* can express their comments in *alt.tv.simpsons*. As you can see, newsgroups are electronic exchanges formed by groups of people interested in sharing information about specific subjects.

The Usenet's Organization

The Usenet is divided into various *hierarchies* that help arrange the thousands of newsgroups into somewhat orderly categories. To this end, each newsgroup contains a brief *prefix* that tells you something about the group. For instance, groups that focus on computing issues typically begin with the letters *comp;* those interested in recreational activities begin with *rec;* and those pertaining to business start with *biz*. Some newsgroups focus on regional issues and begin with appropriate abbreviations. Newsgroups related to issues in Atlanta, for instance, begin with *atl,* and those concerning Chicago begin with *chi*. A complete list of all Usenet prefixes would span several pages, but see Table 3.1 (page 116) for some of the most commonly used prefixes and their meanings.

Following the prefix, the remainder of the newsgroup's name usually provides everything else you need to know about the content of the group. In fact, some newsgroups adopt names with four or five parts to pinpoint their specific focus. For instance, people who keep fish as pets communicate in a group called *rec.aquaria*. Even more specific is a group dedicated only to fish that live in unsalted water, which goes by the longer name of *rec.aquaria.freshwater*. Finally, those who want to discuss certain kinds of freshwater fish can choose such groups as *rec.aquaria.freshwater.cichlids* and *rec.aquaria.freshwater.goldfish*.

TABLE 3.1   Common Newsgroup Prefixes and Meanings

Prefix	Meaning
alt	Covering a wide variety of subjects, almost anyone can establish *alt* groups to discuss almost anything. These groups range from subjects of widespread interest, as in *alt.books,* to those of very limited appeal, as in *alt.pizza. delivery.drivers* or *alt.death-of-superman.*
biz	Groups concerning business topics.
can	Groups concerning Canadian topics.
comp	Groups concerning computer issues.
gov	Groups concerning government issues.
ibm	Groups concerning products and services of the large computer manufacturer IBM.
k-12	Groups concerning primary and secondary education.
Microsoft	Groups hosting discussions concerning products of the world's largest software maker, Microsoft.
misc	A catch-all prefix that applies to newsgroups that often are difficult to classify into other categories.
rec	Groups concerning recreational activities.
sci	Groups concerning scientific issues.

Keep in mind, however, that there's no need to apply or register for the right to read the postings in newsgroups. For example, you don't have to be an accountant to visit *alt.accounting.* And if you just want to visit *alt.food.cooking* once to look for a specific recipe, you can do so even if you never plan to visit the group again.

The Content of Newsgroups

In any discussion of the Usenet, it's essential to convey this warning: *Most newsgroups are uncensored.* They truly represent "free-for-alls," in which anybody can post a message for all to see. While this freedom makes for some lively discussions, it also can bring out the worst in human nature. Unfortunately, some messages—and even some entire newsgroups—consist of racist, sexist, or other offensive ideas. Some are blatantly obscene.

Before you start to explore the Usenet, be aware that (1) such messages are extremely pervasive and (2) you are likely to stumble across some, even in newsgroups that you wouldn't expect to carry such content. It isn't at all unusual to find offensive or adult messages scattered throughout seemingly innocent newsgroups, such as those devoted to sports, hobbies, or professional discussions. Indeed, some adult-oriented businesses splatter their advertisements on virtually every Usenet newsgroup, making such messages difficult to avoid.

A handful of Usenet newsgroups are moderated, meaning that somebody has taken on the difficult job of reading every message computer users attempt to post. Only those messages the moderator deems appropriate actually are posted; the rest are discarded. Some but not all moderated newsgroups contain the word *moderated* in their title, such as *alt.lang.c++.moderated* (for users of the C++ computer programming language).

Accessing the Usenet

You can access the Usenet in a variety of ways. Some web browsers, such as Netscape, contain features that allow users both to visit Usenet newsgroups and post messages to them. A variety of specialized Usenet software also is available for both the PC and Macintosh platforms, including such programs as NewsWatcher and Free Agent. Certain email software programs, such as Pine, are not equipped to read the content of newsgroups or post to them.

In addition, a handful of services are available—such as Deja.com *http://www.deja.com* and Remarq *http://www.remarq.com*—that allow you to access the Usenet through the World Wide Web. These services tend to be more user friendly than stand-alone Usenet software. For instance, they allow you to search through the hundreds of thousands of Usenet messages for specific subjects, rather than forcing you to spend a lot of time browsing through multiple newsgroups looking for the subject you're interested in.

As an example, let's say you're looking for information about how to install cedar siding on your home. Using traditional Usenet software, you would have to browse through the long list of more than 20,000 newsgroups to pick a few in which messages about cedar siding might appear. You might, for instance, select *alt.home.repair* or *alt.building. construction*. Then you'd have to browse through the hundreds of messages in each of those newsgroups to see if anybody had written anything about siding. Using Deja.com or Remarq, however, you simply could type in the words *cedar siding,* and within a few seconds, the service would do the browsing for you. One recent search on *cedar siding* using Deja.com brought up more than 100 Usenet postings, in groups ranging from *alt.building.realestate* to *rec.woodworking.* While this list is still quite long, it's substantially shorter than that produced by searching on the topic with traditional Usenet software.

Deja.com offers another advantage that will limit the number of hits. It advertises that it can help users avoid some of the offensive messages that are posted to the Usenet. This service allows you to screen out so-called adult newsgroups from your Usenet searches. The service claims the screening process eliminates newsgroups with strong sexual content.

At first, accessing the Usenet can be intimidating. With thousands of newsgroups from which to choose and often thousands of messages in each group, it's easy to be overwhelmed. But just as a public library patron doesn't try to read every book in the library, neither should you try to read everything posted on the Usenet. Scan the lists of newsgroups and messages, pick a few topics you're interested in, and begin by browsing just that information. In time, you'll learn which groups are most pertinent to your interests.

The Usefulness of the Usenet

Much of the content of newsgroups concerns avocations such as sports and entertainment. Even so, businesses, government agencies, health care organizations, and educational institutions might also be interested in reading and posting Usenet messages.

Some software manufacturers have created newsgroups dedicated to discussions of their products. Customers can use such groups to ask questions of the company's technical support staff as well as to exchange information. Even software firms that don't have official newsgroups for their products may find customers establishing such groups themselves.

The Usenet is full of groups dedicated to discussing such common software products as Microsoft Windows, WordPerfect, and Netscape. In addition, people in all lines of work have discovered the Usenet to be an effective source of advice and technical information. Many people in the computer field have found solutions to frustrating technical problems by posting questions on the Usenet. Often, others lurking on the Usenet know the answers to these questions and generously respond with answers. So, too, can individuals in nontechnical fields take advantage of the Usenet to solve problems in their work. A farmer can look for advice on eradicating a certain insect, a lawyer can seek tips for drawing up a strategically written legal document, and a professor can search for teaching tips and classroom exercises on multiculturalism. According to the company that runs the Deja.com web site, U.S. government agencies also are using the Usenet. Today, more than 200 specialized newsgroups, designed to disseminate government information, are in use.

In addition, the Usenet has become a popular medium through which employers advertise job opportunities, especially in the computer and technology fields. And some firms have discovered that Usenet postings can help them sell used computer equipment.

Note: While newsgroups generally accept as appropriate "Help wanted" advertisements and "For sale by owner" messages, they do not accept blatant advertisements for products and services. Rather, merchants are advised to find other ways to advertise their wares. We'll talk more about such topics in the section titled Netiquette (pages 122–124).

Writing for the Usenet

Regardless of your purpose for posting messages to the Usenet, it's important to remember that your postings always reflect on either you and/or your company. This means that even though the Usenet is often an informal communication medium, you should still follow some basic rules of professional communication. Do so whether you're using newsgroups for business or personal pursuits.

Many of the communication considerations that apply to email messages—including purpose, audience, style, and tone—also apply to newsgroup messages. There's one difference, however: Newsgroup postings are more public. Think of a newsgroup posting as a form of email that the whole computing world can see.

If you're a new Usenet user, you should visit a newsgroup called *news.announce.newusers* before posting any messages. Designed for novices to the Usenet, this group contains important messages to help you not only navigate through but also post messages to newsgroups. A group of dedicated moderators maintains information concerning Usenet rules of conduct. And as such, it's a must read for every new Usenet visitor. It includes the topics "Welcome to Usenet," "A Primer on How to Work with the Usenet," and "What Is Usenet?"

Even though you should take a look at *news.announce.newusers* to familiarize yourself with the Usenet and Usenet postings in general, you should still spend some time reading postings on various newsgroups before posting any messages of your own. In particular, scanning, or lurking around, the newsgroups you think you may wish to post to will give you a sense of who posts and reads the messages there. For instance, if a newsgroup contains *Microsoft* in its name, is it frequented by computer experts at the company or visited by consumers complaining about (or praising) Microsoft products?

Once you think you're familiar enough with the Usenet to begin posting messages, remember to keep them simple, focused, easy to read, and as short as possible. Remember that your audience will be reading your messages on a computer screen, often inside a window that can only display a few lines at once. Concise prose—both at the paragraph and sentence levels—and blank lines to separate paragraphs will make your messages easier to read and understand. Reader comfort and comprehension are especially important if you're posting a question. Doing so will help guard against off-track responses.

As with email messages, it's best to avoid abbreviations and acronyms in Usenet postings. Remember, your readers come from a wide range of backgrounds, and esoteric abbreviations may hinder (even prevent) them from understanding your message completely. Abbreviations such as *IT, ISP,* and *ISDN* may be common in your industry; however, they may be foreign to your audience. Moreover, part of your audience may quite literally be foreign, reading your message in some other country! So, avoid words or references that may be difficult for people outside your culture, broadly defined, to understand.

This notion of cultural differences brings up another issue related to abbreviations: "Net-icisms"—such as *IMHO* for "in my humble opinion" and *BTW* for "by the way"—are acronyms that replace everyday expressions. While using them may save you some time, they'll surely cost your readers time and may even obscure the meaning of your message. This type of shorthand may be appropriate for informal, personal communication, but it is out of place in more formal business writing.

Of course, you should spell all the words in your message correctly and be sure your grammar is correct. Spell-checker and grammar-checker software can help if you aren't confident of your own ability. But don't allow these aids to substitute for proofreading the old-fashioned way.

One other word of caution: DON'T TYPE YOUR MESSAGES IN ALL UPPERCASE LETTERS. Besides being difficult to read, it's rude. Instead, use upper- and lowercase letters. It's easier to read and more polite.

> *Note:* Most of the effective electronic communication strategies—including those regarding purpose, audience, style, and tone—that apply to email messages and web sites also apply to newsgroup postings, so please review those sections in the email and web chapters.

Your Subject Line

When composing a Usenet message, choose your subject line carefully. Keep in mind that when readers log onto a particular newsgroup, their software will display a long list of subject lines, one for each message posted since the reader last visited that group. Your word choice for the subject line will be the sole factor by which a computer user decides whether to read your message.

First, pick a brief subject that accurately conveys the content of your message. Generic titles such as "Important message," "Please read this," and "I'm back" are too vague to let your reader know the focus of your message. Instead, pick something more specific, such as "How do I file IRS form 8063?" or "Jobs available for experienced C++ programmers." These simple strategies should help draw appropriate attention to your postings.

Where to Post Your Message

Exercise some care in deciding which newsgroups are most appropriate for your messages. A question related to portable computers, for instance, would likely receive a large number of responses when posted to a group that's dedicated to such questions, *comp.sys.laptops*. It may receive fewer responses from a more general computing group, such as *comp.answers*. Select the one or two groups that seem most appropriate for your message. It's considered impolite to post the same message to multiple newsgroups.

Consider also whether regional factors might affect the newsgroup you choose to post your message to. For instance, if you are trying to sell your car, stick with newsgroups of local interest. If your car is in Kalamazoo, no one in Madrid will come to look at it. Clearly, then, there's no need to post a car ad for the whole world to see.

Before posting a *question* to any newsgroup, make sure the same question hasn't already been posed by another user. On many newsgroups, you'll find a posting titled *FAQs,* for "Frequently Asked Questions," which may already contain the answer you seek.

Netiquette

Postings on the Usenet often grow into long discussions, or *threads,* with dozens of users expressing their opinions on a given subject. For instance, suppose you post a message on *comp.sys.laptops,* asking for advice on which portable computer to purchase for your job, which involves traveling. You may get a few responses advocating IBM products, others from people who swear by Compaqs, and still others who recommend Hewlett-Packard. Some people may pass along personal anecdotes about why they endorse one brand over another, and you may even get some responses from employees of the various computer makers, who may attempt to answer questions about their products. Likewise, long discussions can break out in movie newsgroups over the pros and cons of various flicks, in medical newsgroups over treatment options for various diseases, and in political newsgroups over the attributes of certain candidates.

These Usenet discussions have their own rules of etiquette, or *netiquette.* Before entering such a discussion, read what already has been written. Often, when one person poses a question on the Usenet, he or she gets flooded with dozens of identical responses. If somebody else has already said what you planned to say, there's no need for you to post a similar message. And by all means, avoid posting simple affirmations to earlier responses. Messages such as "Me too" and "I agree" don't advance a particular discussion or argument. Besides, newsgroup participants don't know who you are. Accordingly, your status, position, and character are low, so your "Me too" will typically mean nothing. If you have nothing *substantial* to add to any discussion, refrain from participating in it.

Netiquette also suggests that when responding to an earlier message, you quote part of the original message. That way, readers who didn't have a chance to see the earlier message can understand the context of your remarks. However, don't overdo it with quotations. They can make your messages longer than necessary. Simply pick out one or two relevant lines in the earlier message, quote them, and respond to them. Quoting long blocks of copy wastes disk space and lengthens the time it takes other people's computers to download your message.

If you are responding to an earlier post, consider whether it makes more sense to send that response through the Usenet for all to see or through email directly to the poster of the original message. Choosing email means that only the person who posted the first message will see your response. This can be an advantage if you're communicating about an issue that lacks widespread interest—for instance, a very specific question about a technical computer issue.

While the majority of communication on the Usenet is polite, some writers try to turn the medium into a forum for confrontation. They get insulted if somebody disagrees with their views, and they insult those with whom they disagree. Anyone who regularly browses the Usenet almost certainly has encountered people like this. A simple discussion over, say, the advantages and disadvantages of a particular brand of software may suddenly degenerate into electronic name-calling, with supporters of one product squaring off against those who prefer another.

The practice of sending these uncivil messages is called *flaming,* and it should never be a part of business or personal communication. For example, consider a recent discussion on a newsgroup dedicated to the subject of rock climbing. One writer posted a message to the group inquiring about the death of a well-known climber, who was killed while attempting a dangerous jump. But the writer made the mistake of spelling the climber's name wrong, setting off a string of nasty flames from other posters. "Learn to spell his name correctly!" admonished one person. "Do your homework before asking a thousand people a question!" scolded another. Ultimately, even the original poster was drawn into the verbal battle. "Now that's a warm reception," he wrote in a sarcastic reply. "Thank you for the incredibly useful information, oh spelling wizard." Needless to say, the verbal barbs served only to take the conversation off track and provided no answers to the original question.

We are all free to express our opinions on the Usenet. However, we also must remember that we are dealing with *people* on the other end of our computer wires. They have feelings and deserve to be treated with respect.

The Global Usenet Audience

The content of all your Usenet messages should be just as carefully written as a paper memo you would send to the president of your company or university. Why? Because the president—and everyone else at your firm or college—can read what you've written, even if you didn't intend for them to be in the audience. Remember, virtually anyone who owns a computer can access and read Usenet messages—that means your present and future employers, your present and future employees, your competitors, your spouse, your exspouse, your pastor, your neighbors—in other words, everyone! Furthermore, those readers can easily print and download your message to their computer hard drive, where it can live on forever. And if you post your message from your business Internet account, there's a good chance the name of your employer will appear somewhere in the message, such as in your "Reply to" address. Some readers may even assume that you're speaking for your company, even if you're posting only your own opinions.

In some ways, posting a message to the Usenet is akin to publishing a comment for the world to see. As such, it's theoretically possible to commit libel through a Usenet posting. While this is an uncharted area of the law, it's prudent for you to follow the same precautions when posting a message to the Usenet that you would if you were publishing a brochure or writing a newspaper column. That is, carefully check your facts, attribute material gleaned from other sources, and express your opinions in an ethical, responsible manner. The same ethical rules discourage posting information that advocates or gives instructions for performing illegal acts, such as building bombs, soliciting prostitutes, and pirating cable television signals.

If you are posting material on the Usenet that was originally written by somebody else—such as a newspaper article, song lyrics, or an excerpt from an instruction manual—be sure to look into whether the material is copyrighted. Seek the advice of a qualified attorney before posting material that may be subject to copyright laws.

At its best, the Usenet can serve as something of a "town square" on the Internet—a place where individuals from different countries and of different backgrounds can share opinions and help one another. Following simple rules of professional discourse will help ensure that your experience with the Usenet will be productive.

Summary

Overview

- The Usenet consists of more than 20,000 newsgroups, each pertaining to a specific subject.

- In most newsgroups, anybody is welcome to post and respond to messages.

- An estimated 100,000 messages are posted each day.

- The Usenet can be accessed through newsreader software, such as NewsWatcher and Free Agent, or the "News" window of Netscape Navigator.

- The Usenet also can be accessed through the World Wide Web on such sites as *www.deja.com.*

- Newsgroups are divided into hierarchies, which are designated by the first few letters of the newsgroup name—for instance, *comp., rec.,* and *alt.*

Newsgroup Subjects

- Newsgroup subjects range from vocational to recreational to professional.

- Each newsgroup's name tells you something about its subject.

- Businesses are increasingly taking advantage of the Usenet to exchange information with customers and advertise job opportunities.

- People in some professions, such as computer programming and accounting, use newsgroups widely to share information regarding their occupations.

- Most newsgroups are uncensored, and many contain offensive or obscene content, often mixed in with the regular content.

- Because posting a message on the Usenet is somewhat akin to publishing, you should make sure your postings do not run afoul of libel or copyright law.

Appropriate Style and Tone

- Postings to the Usenet should follow many of the same conventions that email does.

- Messages should be short, direct, and mindful of their audiences.

- Readers will judge you and your company by the quality and content of your Usenet posts.

- Usenet messages should have accurate and informative "Subject" lines.

- Post messages to newsgroups that pertain to the appropriate subject matter of the message.

- It's considered a violation of Usenet etiquette to post the same message to many different newsgroups; rather, pick the one or two groups likely to generate the most responses.

- Responses to messages posted by others should be polite, civil, and aimed at moving the online conversation forward.

- Flaming—the practice of sending mean or insulting replies—is discouraged.

- Posting a response that merely affirms other earlier responses is also discouraged.

Index

writing guidelines *continued*
   organization, 29–33, 79–81,
      115–116
   paragraphs, 87–88
   punctuation, 51–52
   redundancies, 77
   sentences, 87–88
   simple vs. overblown vocabulary,
      75–77

spelling, 82–83, 121
style and tone. *See* style and tone
Usenet, 119–124
web sites, 68–81, 119–120
word choice, 71–79
WWW. *See* World Wide Web